Engineering Materials

Springer
*Berlin
Heidelberg
New York
Barcelona
Hong Kong
London
Milan
Paris
Tokyo*

Robert Kratz · Peter Wyder

Principles of Pulsed Magnet Design

With 110 Figures and 29 Tables

 Springer

Dr. Robert Kratz
General Atomics
3550 General Atomics Court, San Diego, CA 92121-1122, USA
e-mail: Robert.Kratz@gat.com

Professor Dr. Peter Wyder
Hochfeld-Magnetlabor Grenoble
Avenue des Martyrs (C.N.R.S.), B.P. 166, F-38042 Grenoble Cedex 9, France
e-mail: wyder@labs.polycnrs-gre.fr

Library of Congress Cataloging-in-Publication Data.

Kratz, Robert, 1961-
Principles of pulsed magnet design / Robert Kratz, Peter Wyder. p. cm. – (Engineering materials)
Includes bibliographical references and index.
ISBN 3540437010 (alk. paper)
1. Electromagnets–Design and construction. I. Wyder, Peter, 1934- II. Title. III. Series.

QC760 .K686 2002
621.34–dc21   2002070590

ISBN 3-540-43701-0 Springer-Verlag Berlin Heidelberg New York

This work is subject to copyright. All rights are reserved, whether the whole or part of the material is concerned, specifically the rights of translation, reprinting, reuse of illustrations, recitation, broadcasting, reproduction on microfilm or in any other way, and storage in data banks. Duplication of this publication or parts thereof is permitted only under the provisions of the German Copyright Law of September 9, 1965, in its current version, and permission for use must always be obtained from Springer-Verlag. Violations are liable for prosecution under the German Copyright Law.

Springer-Verlag Berlin Heidelberg New York a member of BertelsmannSpringer
Science + Business Media GmbH

http://www.springer.de

© Springer-Verlag Berlin Heidelberg 2002
Printed in Germany

The use of general descriptive names, registered names, trademarks, etc. in this publication does not imply, even in the absence of a specific statement, that such names are exempt from the relevant protective laws and regulations and therefore free for general use.

Typesetting: Camera-ready from the authors using a Springer LaTeX 2_ε macro package
Cover design: de'blik, Berlin

Printed on acid-free paper   SPIN: 10880169   57/3141 - 5 4 3 2 1 0

Preface

This book deals with the design of pulsed coils for the generation of high magnetic fields. The scope is limited to nondestructive coils. The book's purpose is to provide the designer of a pulsed field facility, the inquisitive student and the scientist with a concise and comprehensive text which describes every aspect of coil design. Special emphasis is laid upon first-order principles, which allow an estimation with paper and pencil and are important for an understanding of the basic design principles. These design formulas are then supplemented by numerical calculations and simulations. The physics necessary to describe a pulsed coil are the theories of electromagnetism, of continuum mechanics and of thermodynamics. It is the combination of these fields of physics which make the construction of a coil at first sight seem a rather tedious process. In this book we want to lead any newcomer to the club through this jungle and we describe the meaning of all possible design variables. The path to the final construction starts with only the most important of these design variables, and later the finer details of 'second order' get incorporated.

We have divided the book into four chapters. In Chapter 1 we give an introduction to the necessary physics. We give an overview of several methods for calculating the magnetic field of a coil and describe the law of Biot-Savart and introduce the scalar magnetic potential and the magnetic vector potential. Furthermore we introduce the inductance of a coil and finally apply methods we have developed to solenoids. The chapter dealing with continuum mechanics introduces the stress and strain tensors and derives the equation for the equilibrium of forces in a rigid body, which relates the external forces to the intrinsic stresses in the body. This equation is completed with the so-called stress-strain relations. Again, after the theoretical description we describe the situation for a solenoid. Finally, we write down the equations necessary to predict the heating of the coil. In a pulsed coil this heating process can be regarded as adiabatic. The heating effects we consider are ohmic heating and heating due to the magnetoresistance. The data for a few conductive materials are presented and the superiority of copper, if only heating from liquid nitrogen to room temperature is considered, as a material for pulsed magnets is demonstrated.

Chapter 2 discusses the analytical calculations of coils. We follow here a somewhat historical path and concentrate first on the optimal use of the available power in a solenoid. With increasingly sophisticated current distributions the power efficiency can be increased. The coil with constant current density, which is easy to manufacture, is surpassed in efficiency by the Bitter coil and the Gaume coil, and the best power efficiency is achieved with the so-called Kelvin distribution. Even with the optimal current density the generation of ever-higher fields needs more and more power, and eventually there occur two limitations, namely the build-up of huge Lorentz forces and the finite cooling capability. For an optimized coil both limitations lead to a current density different from the Kelvin distribution. The limited cooling capability does not occur if one goes over to pulsed magnet systems, where the field generating current flows only for a short time interval, during which the heat capacity of the coil serves as a reservoir. The optimization of real coils with respect to the dissipated power can be performed in an analytical way. This is not so however, for the mechanical stresses, which means that a simplified model becomes mandatory. For this purpose we treat the coil as extended to infinity in the axial direction. The axial field as well as the stresses can then be calculated by hand. We describe this technique for several coil types, for instance for polyhelix coils, wire-wound coils and two-coil systems. Additionally, we investigate various distributions of the current density such as the constant current density, the $1/r$-distribution of the Bitter coil or a current density distribution resulting in constant stresses. A concluding section deals with eddy currents and we derive estimations for a closed cylinder and a cylinder with a slit, both being important geometries in a solenoid.

Chapter 3 deals with numerical simulations and calculations. It is the numerical equivalent of the analytical calculations of Chapter 2. Besides being one step closer to reality, the comparison with the analytical estimations provides a check on the quality of such estimations. Coils of the polyhelix type are simulated with a standard optimization algorithm, and the calculation of wire-wound coils is performed by the method of finite elements.

In Chaper 4 we describe pulsed field facilities, by which we mean the complete system of energy source and high-field coil. Depending on the type of the energy source, the equations for the system describe either capacitive or inductive energy storage systems. We describe capacitive energy storage and begin with the simplest case of constant circuit elements, which we then refine gradually with the incorporation of the heating of the high-field coil, the magnetoresistance and finally eddy currents in the coil. For inductive energy storage we restrict the discussion to constant coil parameters, but deal more extensively with possible mutual inductances between the inductive storage coil and the high-field coil. Furthermore, we discuss how to increase the energy transfer for the inductive storage method. We end with an overview of several possible energy sources.

San Diego, Grenoble *Robert Kratz*
May 2002 *Peter Wyder*

Contents

1. **Basics** 1
   1.1 Calculation of Magnetic Fields 2
      1.1.1 Maxwell's Equations 3
      1.1.2 Lorentz Force 4
      1.1.3 Static and Pulsed Coils 4
      1.1.4 Solutions of $j = \nabla \times H$ 5
      1.1.5 The Law of Biot-Savart 6
      1.1.6 The Inductance 7
      1.1.7 Application to Solenoids 9
   1.2 Forces and Stresses 21
      1.2.1 Elasticity 21
      1.2.2 Stress and Strain 22
      1.2.3 Equilibrium of Forces 24
      1.2.4 Stress-Strain Relations 25
      1.2.5 Application to a Solenoid 25
      1.2.6 Strength Theories 26
   1.3 Heating of Pulsed Coils 27
      1.3.1 Heating Process 27
      1.3.2 Material Selection 29
      1.3.3 Data for Conductors 31
      1.3.4 Magnetoresistance 33

2. **Analytical Calculations** 35
   2.1 Classical Solenoids 36
      2.1.1 A Ring-Shaped Current Element 37
      2.1.2 Calculation of the Fabry Factor 40
      2.1.3 Cylindrical Coil with Constant Current Density ... 42
      2.1.4 Coil with Current Density $\sim 1/r$ 43
      2.1.5 The Gaume Current Distribution 44
      2.1.6 The Kelvin Current Distribution 45
      2.1.7 The Optimal Magnet with Kelvin Distribution 46
      2.1.8 Coils with Optimal Shape 47
      2.1.9 Conclusion 48
   2.2 No Radial Transmission of Forces 54

		2.2.1	Boundary Conditions	57
		2.2.2	Coil with Constant Current Density	59
		2.2.3	Bitter Coil	60
		2.2.4	Stress-Optimized Coil	61
		2.2.5	Comparison of the Three Coil Types	66
		2.2.6	Examples for Pulsed Coils	71
	2.3	Radial Transmission of Forces		73
		2.3.1	Coil with Constant Current Density	73
		2.3.2	Bitter Coil	81
3.	**Numerical Simulations**			**89**
	3.1	Polyhelix Coils		89
		3.1.1	Coil Model	89
		3.1.2	Optimization with Constraints	92
		3.1.3	The Optimal Current Density	93
		3.1.4	Variation of the Stored Magnetic Energy	96
		3.1.5	Variation of Coil Size and Material Strength	98
		3.1.6	Coil with Smaller Inner Radius	100
		3.1.7	Comparison with Analytical Methods	100
		3.1.8	Estimations and Reality	104
	3.2	Wire-Wound Coils		106
		3.2.1	Calculation of a Coil	107
		3.2.2	'Search' for an Optimal Coil	113
		3.2.3	Comparison with Analytical Methods	116
		3.2.4	Two-Coil System	119
	3.3	Plastic Deformation		123
		3.3.1	Zylon Reinforcement	125
		3.3.2	S2-Glass Reinforcement	127
		3.3.3	Conclusion	128
4.	**Pulsed Field Facilities**			**131**
	4.1	Capacitive Energy Storage		132
		4.1.1	Ideal RLC Circuit	132
		4.1.2	Resistance Increase Due to Heating	134
		4.1.3	Magnetoresistance	136
		4.1.4	Eddy Currents	140
		4.1.5	RLC Circuit with Crowbar Diode	145
	4.2	Inductive Energy Transfer		147
		4.2.1	Ideal Coils with No Coupling	147
		4.2.2	Ideal Coils with Coupling	150
		4.2.3	Real Coils with Coupling	154
		4.2.4	Improvements	155
	4.3	Energy Sources		159
		4.3.1	Local Power Grid	161
		4.3.2	AC Generator	162

	4.3.3	Battery	163
	4.3.4	Homopolar Generator	163
	4.3.5	Inductive Energy Storage	165
	4.3.6	Capacitor Bank	165
	4.3.7	Conclusion	166

References 169

Index 179

1. Basics

In this chapter we introduce the most fundamental principles which one encounters when designing a coil for the generation of magnetic fields. We restrict the discussion to 'first-order effects', so that a reader new to the field of pulsed magnets can see the basics clearly.

The starting point for the calculation of the fields of a coil are Maxwell's equations. We immediately restrict ourself to the stationary and the quasi-stationary case. Stationary means we allow constant currents to flow, but neglect the effects of switching the currents on and off. Quasi-stationary means that we allow the current to vary with time, but neglect any feedback onto the driving currents such as, for instance, eddy currents. Additionally, there are no ferromagnetic materials involved. Within this framework we develop methods for calculating the magnetic field, the scalar magnetic potential and the vector potential, as well as the inductance and the magnetic energy of a solenoid.

The magnetic field and the current flowing through the coil give rise to Lorentz forces, which must be compensated by mechanical forces within the coil; the governing equation is called the equilibrium condition. For the calculation of the mechanical stresses we apply the theory of elasticity, but restrict ourselves to the isotropic case. We introduce the tensors of stress and strain, and the concept of the von Mises stress, being one of the invariants of the stress tensor. The von Mises stress maps a stress tensor to a scalar value and is often used as criterion for deciding whether a material is in the elastic or the plastic regime. The equilibrium conditions are simplified for the case of cylindrical symmetry in a solenoid, and even more so for the case of 'free-standing turns' in the coil.

Finally, we write a few words about the heating of pulsed coils. Because of the adiabatic heating process – during the current pulse of a few milliseconds there is practically no heat transfer from the coil to its environment – the development of the heating equation is straightforward. After that we concentrate on the selection of the best conductor material and the proper temperature range for the initial and final temperatures of the coil. Among the effects that we shall neglect are magnetoresistance, eddy currents, plastic deformation and thermal stresses. The magnetoresistance describes the increase of the resistivity of a conductor material in a magnetic field. The increased resistivity leads to an increase in the heating of the coil, which causes the final temperature of the coil to be reached in a shorter time; the effects of magnetoresistance decrease the pulse length. In a coil with constant current density the magnetoresistance leads to a temperature distribution within the coil, which then causes thermally induced mechanical stresses. A temperature distribution and thermal stresses occur also in coils with non-constant current densities. Eddy currents try to expel the current from the inside of a conductor. This causes a time-dependent fine structure of the current density. The effect on the main magnetic field at the center may be really negligible, but eddy currents lead to an increased heating of the conductor at its surface. Furthermore, the inductance of the coil becomes a function of time.

1.1 Calculation of Magnetic Fields

In the art of building coils generating extremely high magnetic fields the knowledge of the magnetic field at every point in space is of fundamental importance. From the view of an experimenter the field at the center of a probe volume as well as the field quality, i.e. the deviation of the magnetic field over this volume with respect to the field at the center, have to be known.

For the designer of a magnet the fields in the coil volume and hence the Lorentz forces are of greatest interest. In high-field coils the Lorentz forces acting on the windings are quite huge and an inappropriately constructed magnet would literally destroy itself. Therefore, the fields have to be known in advance and a way has to be found to calculate the fields produced by an arbitrary coil.

Other quantities one often wants to know are the mutual inductances or forces acting between two coils. These may be small pickup coils or a second magnet in a system of two or more coils, as for instance in a typical NMR magnet. These forces may be deducted from the mutual inductances.

Finally the calculation of the field is essential in so-called inverse problems. In this case one starts by demanding a certain field intensity and a certain field quality within a probe volume and tries to construct the magnet system which just has these properties. Usually in these sort of problems additional constraints such as minimal weight, minimal power consumption, a minimum bore, etc. are incorporated. Typical examples are NMR magnets, which require a very high field homogeneity, or special coils for the levitation of water (diamagnetic levitation), which require strong fields combined with strong field gradients [1–5].

We start with a short course on electrodynamics and begin with Maxwell's equations [6, 7].

1.1.1 Maxwell's Equations

The axioms of electrodynamics are summarized in four elegant equations, Maxwell's equations (J.C. Maxwell, 1864). In integral form they are:

$$\frac{d}{dt}\int \boldsymbol{B} \cdot d\boldsymbol{F} = -\int \boldsymbol{E} \cdot d\boldsymbol{s}, \tag{1.1}$$

$$\int \left(\frac{\partial}{\partial t}\boldsymbol{D} + \boldsymbol{j}\right) \cdot d\boldsymbol{F} = \int \boldsymbol{H} \cdot d\boldsymbol{s}, \tag{1.2}$$

$$\int \boldsymbol{B} \cdot d\boldsymbol{F} = 0, \tag{1.3}$$

$$\int \boldsymbol{D} \cdot d\boldsymbol{F} = \int \varrho \, dV. \tag{1.4}$$

Here we call $\boldsymbol{B}$ the magnetic field strength and $\boldsymbol{E}$ the electric field strength. $\boldsymbol{D}$ is known as the displacement, and we refer to $\boldsymbol{H}$ as the H-field. Finally, $\boldsymbol{j}$ is the current density and ϱ the charge density.[1]

The first equation describes Faraday's law of electromagnetic induction, which relates the magnetic flux $\int \boldsymbol{B} \cdot d\boldsymbol{F}$ through a surface F to the line integral of the electric field $\boldsymbol{E}$ over the boundary of that surface F. Faraday's law states that the change of the magnetic flux through a surface is proportional to the negative ring voltage $\int \boldsymbol{E} \cdot d\boldsymbol{s}$ along the boundary. The negative sign is an expression of Lenz's law.

The second equation is Ampere's law in its modified form. It connects the current flowing through a surface F with the magnetic ring voltage $\int \boldsymbol{H} \cdot d\boldsymbol{s}$ on the boundary of that surface F. The current through the surface F consists of the electric current $\int \boldsymbol{j} \cdot d\boldsymbol{F}$ and the displacement current $\int \frac{\partial}{\partial t}\boldsymbol{D} \cdot d\boldsymbol{F}$.

The last two equations give relationships between properties in a volume V and on the surface F of that volume. The third equation shows that the magnetic flux through such a closed surface F is zero, which means the $\boldsymbol{B}$-field is source-free.

The fourth equation is Gauss's law, the total charge in the volume, $\int \varrho \, dV$, is equal to the integral of the displacement $\boldsymbol{D}$ over the surface F of the volume.

The differential form of Maxwell's equations is written as

$$\frac{\partial}{\partial t}\boldsymbol{B} = -\nabla \times \boldsymbol{E}, \tag{1.5}$$

$$\frac{\partial}{\partial t}\boldsymbol{D} + \boldsymbol{j} = \nabla \times \boldsymbol{H}, \tag{1.6}$$

$$\nabla \cdot \boldsymbol{B} = 0, \tag{1.7}$$

$$\nabla \cdot \boldsymbol{D} = \varrho. \tag{1.8}$$

---

[1] Unfortunately very often $\boldsymbol{B}$ is called the magnetic induction and $\boldsymbol{H}$ the magnetic field strength. For an overview of the controversy between $\boldsymbol{B}$ and $\boldsymbol{H}$ see [8,9].

Equations (1.5–1.8) form a linear, partial differential equation system. A consequence of the linearity is the superposition principle. The equations are not sufficient to solve for the five vectors $\boldsymbol{E}$, $\boldsymbol{B}$, $\boldsymbol{D}$, $\boldsymbol{H}$ and $\boldsymbol{j}$, however. We need additional properties giving some more relations between the vectors. These material constants are known as the conductivity σ, the permittivity ε and the permeability μ:

$$\boldsymbol{j} = \sigma\,\boldsymbol{E}\,, \tag{1.9}$$
$$\boldsymbol{D} = \varepsilon\,\boldsymbol{E}\,, \tag{1.10}$$
$$\boldsymbol{B} = \mu\,\boldsymbol{H}\,. \tag{1.11}$$

In general these three material constants are tensors of rank 2; however, in the isotropic case they are reduced to scalar quantities.

1.1.2 Lorentz Force

The force on an electric charge q depends not only on where it is, but also on how fast it is moving. First there is the electric force, which gives a force component independent of the motion of the charge. We describe it by the electric field $\boldsymbol{E}$. Second, there is an additional force component, the magnetic force, which depends on the velocity of the charge and the magnetic field $\boldsymbol{B}$. The total force on the particle, also called the Lorentz force $\boldsymbol{F}$, can be written as

$$\boldsymbol{F} = q(\boldsymbol{E} + \boldsymbol{v} \times \boldsymbol{B})\,. \tag{1.12}$$

If instead of a fixed charge we consider an extended charge density ϱ, we must replace all quantities in (1.12) with their respective counterparts. We find for the density $\boldsymbol{f}$ of the Lorentz force:

$$\boldsymbol{f} = \varrho\,\boldsymbol{E} + \boldsymbol{j} \times \boldsymbol{B}\,. \tag{1.13}$$

1.1.3 Static and Pulsed Coils

So far we have introduced the most general case of the electromagnetic field, described through Maxwell's equations. In the case of static coils the full Maxwell equations are simplified, and we have the so-called magnetostatic case with stationary fields. All time derivatives disappear and the second of Maxwell's equations becomes:

$$\boldsymbol{j} = \nabla \times \boldsymbol{H}\,. \tag{1.14}$$

In the case of quasi-stationary fields we calculate the fields as in the stationary case and assume time dependences only in the first order. Mathematically speaking, we split the current density into a spatial and a time-dependent part:

$$\boldsymbol{j}(\boldsymbol{r},t) = \boldsymbol{j}_r(\boldsymbol{r}) \cdot g(t) \,, \tag{1.15}$$

and solve (1.14) for $\boldsymbol{j}_r(\boldsymbol{r})$.

A pulsed coil is a typical example of a quasi-stationary field. The field at every moment of time is calculated as if it is the static field of the coil with a certain current density. The magnetic field follows the same function of time as the current density. The real situation, namely the feedback of the changing magnetic field onto the electric field ($\frac{\partial}{\partial t}\boldsymbol{B} = -\nabla \times \boldsymbol{E}$) and therefore also on the current density, is neglected in the quasi-stationary model. Of course the timescale in question determines whether the quasi-stationary model is applicable or whether one has to turn to the full set of Maxwell's equations.

1.1.4 Solutions of $\boldsymbol{j} = \nabla \times \boldsymbol{H}$

For calculating the magnetic field of a coil one has to solve the equation $\boldsymbol{j} = \nabla \times \boldsymbol{H}$. Since the current density $\boldsymbol{j}$ disappears outside of the coil, one can discern between methods solving for the field only outside the coil and methods solving for the field also inside the coil. The first method leads to the introduction of a scalar magnetic potential, the second to the magnetic vector potential.

The Scalar Magnetic Potential ψ. Outside of conductors the equation $\nabla \times \boldsymbol{H} = 0$ is solved by a scalar magnetic potential ψ which is introduced as

$$\boldsymbol{H} = -\nabla \psi \,. \tag{1.16}$$

If we further assume a constant permeability[2] $\mu = \mu_0$, we can use the condition $\nabla \cdot \boldsymbol{B} = 0$ and get Laplace's equation:

$$\Delta \psi = 0 \,. \tag{1.17}$$

The potential ψ is not unique, however, as we can immediately see from (1.16): any constant added to the potential ψ leaves the field $\boldsymbol{H}$ unaffected. Consequently, the integration of (1.16) depends on the integration path. If the integration path is a closed loop surrounding a conductor with current I flowing through it, we get for the line integral of $\boldsymbol{H}$:

$$\psi_1 - \psi_2 = \int \boldsymbol{H} \cdot \mathrm{d}\boldsymbol{s} = \pm I \,. \tag{1.18}$$

The sign depends on how we run around the current (clockwise versus counterclockwise), ψ_1 and ψ_2 denote the potential at the start and the end point of the integration path. In order to make the scalar potential ψ unique we have therefore to forbid integration paths which surround any currents. This is the case if we restrict the region where ψ is defined to a single connected region.

---

[2] $\mu_0 = 4\pi\, 10^{-7} \mathrm{Hm}^{-1}$ is the permeability of free space

1. Basics

The Magnetic Vector Potential A. The magnetic vector potential A allows us to calculate the magnetic field everywhere in space. The starting point is the source-free condition for the magnetic field, $\nabla \cdot B = 0$, which is always fulfilled if B is written as

$$B = \nabla \times A . \tag{1.19}$$

The vector potential A is not unique but is defined only up to the gradient of a scalar function f, because $\nabla \times (\nabla f) \equiv 0$. If A yields the magnetic field B, then $A + \nabla f$ is also a solution. Choosing a certain ∇f is called a 'gauge transformation'.

In the case of a constant permeability $\mu = \mu_0$ we get for (1.14)

$$\Delta A - \nabla (\nabla \cdot A) = -\mu_0 j . \tag{1.20}$$

By choosing a certain scalar function f we can now force the divergence of A to disappear. This so-called gauge condition makes the vector potential unique, it is called the Coulomb gauge.

$$\nabla \cdot A = 0 , \quad \text{Coulomb gauge} . \tag{1.21}$$

Hence (1.20) reduces to

$$\Delta A = -\mu_0 j , \tag{1.22}$$

which has the solution

$$A(r) = \frac{\mu_0}{4\pi} \int_{V'} \frac{j(r')}{|r - r'|} dV' . \tag{1.23}$$

This is the vector potential of a stationary current distribution with density $j(r')$. The current density is limited to a volume V'. Additionally we have assumed here a constant permeability of $\mu = \mu_0$. One can also show that the gauge condition (1.21) is also fulfilled.

1.1.5 The Law of Biot-Savart

In (1.23) we consider a small volume element dV' around the coordinate r', which generates a vector potential dA at the coordinate r of

$$dA(r) = \frac{\mu_0}{4\pi} \frac{j(r')}{|r - r'|} dV' . \tag{1.24}$$

The situation is sketched in Fig. 1.1. By applying the rotation we get the magnetic field dB:

$$dB(r) = \nabla \times dA(r) = \frac{\mu_0}{4\pi} \frac{j(r') \times (r - r')}{|r - r'|^3} dV' . \tag{1.25}$$

Equation (1.25) is called the Biot-Savart law.

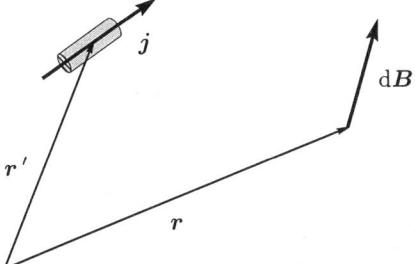

Fig. 1.1. Situation for the law of Biot-Savart. An infinitesimal current element at the position r' with current density j generates a magnetic field dB at the position r

1.1.6 The Inductance

We calculate the magnetic energy W_m of a system of two coils (indices 1 and 2; see also Fig. 1.2) and integrate the magnetic energy density over the whole space:

$$W_\mathrm{m} = \frac{1}{2}\int \boldsymbol{B}\cdot\boldsymbol{H}\,\mathrm{d}V. \tag{1.26}$$

Introducing the vector potential (1.23) we find

$$W_\mathrm{m} = \frac{1}{2}\int \boldsymbol{A}\cdot\boldsymbol{j}\,\mathrm{d}V, \tag{1.27}$$

where the integration is carried out over the volume of the two coils, because only there does the current density not disappear. Using (1.23) a second time we get

$$W_\mathrm{m} = \frac{\mu_0}{8\pi}\iint \frac{\boldsymbol{j}_i\cdot\boldsymbol{j}_k}{|\boldsymbol{r}_i-\boldsymbol{r}'_k|}\,\mathrm{d}V_i\,\mathrm{d}V'_k \tag{1.28}$$

and we have to discern between four cases for the indices i and k:

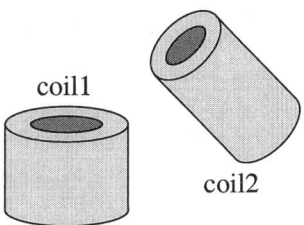

Fig. 1.2. Calculation of the magnetic energy for an assembly of two coils. The starting point is the integration of the magnetic energy density over the whole of space

(a) $i = 1$ and $k = 1$,
(b) $i = 1$ and $k = 2$,
(c) $i = 2$ and $k = 1$,
(d) $i = 2$ and $k = 2$.

Because of the symmetry of (1.28) the cases (b) and (c) are identical. We write the result in the form

$$W_m = \frac{1}{2}\left(L_{11} I_1^2 + L_{22} I_2^2 + 2 L_{12} I_1 I_2\right), \tag{1.29}$$

where I_1 and I_2 are the total current through the cross-section of coil 1 and coil 2, respectively,

$$I_n = \int \boldsymbol{j}_n \cdot \mathrm{d}\boldsymbol{F}_n \quad \text{and} \quad n = 1, 2, \tag{1.30}$$

and the terms L_{ik} are called the inductances. Furthermore, L_{11} and L_{22} are the self-inductances and L_{12} is referred to as the mutual inductance. They are defined as

$$L_{ik} = \frac{\mu_0}{4\pi} \frac{1}{I_i I_k} \iint \frac{\boldsymbol{j}_i \cdot \boldsymbol{j}_k}{|\boldsymbol{r}_i - \boldsymbol{r}'_k|} \, \mathrm{d}V_i \, \mathrm{d}V'_k, \tag{1.31}$$

or, if we introduce the vector potential again,

$$L_{ik} = \frac{1}{I_i I_k} \int \boldsymbol{j}_i \cdot \boldsymbol{A}_k \, \mathrm{d}V_i. \tag{1.32}$$

In words: the inductance L_{ik} between two conductors i and k is found by the integration of the scalar product of the current density $\boldsymbol{j}_i$ of the conductor i and the vector potential $\boldsymbol{A}_k$ from the conductor k. The volume of integration spans over the whole conductor i. Division by the currents through conductor i and k gives the inductance.

The drawback of this general method is that in general the vector potential cannot be given in closed form, but is itself found only by numerical means. Therefore, numerous techniques and formulas have been developed for calculating the inductance of a conductor system. Before the advent of modern computers these techniques where the only way to calculate an inductance. Even today these methods are widely used, first because they are fast and second because they enable us to see trends for slight changes of geometry. Furthermore, these methods can be fairly accurate. The most complete reference is the book by Grover [10]; other references are found in [11–24].

Finally, we derive an expression for the force between two coils by calculating the negative gradient of the magnetic energy of the two coils, (1.29). The force is then

$$\boldsymbol{F} = -\nabla W_m = -I_1 I_2 \nabla L_{12}. \tag{1.33}$$

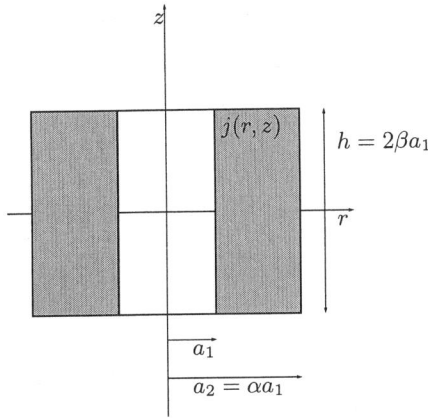

Fig. 1.3. Cylindrical coil with rectangular cross-section and inner radius a_1. The outer radius and height are $a_2 = \alpha\, a_1$ and $h = 2\beta\, a_1$, respectively. The current density j within the cross-section depends only on the radial and/or the axial position, but not on the azimuthal position (cylindrical coordinates)

1.1.7 Application to Solenoids

We apply the formalism developed so far to coils with cylindrical symmetry, i.e. solenoids, (see Fig. 1.3), and demonstrate how to calculate the scalar potential, the vector potential and the magnetic field for two representative coil types. These are the coil with constant current density throughout its cross-section, which is approximately realized with a wire-wound coil. The second coil type is known as the Bitter coil, with a current density proportional to $1/r$. In both cases the approximation is that any insulation of the conductor or cooling channels is neglected. Finally, we represent graphs for calculating the self-inductance for the two coil types.

The Scalar Magnetic Potential ψ. The method with the scalar magnetic potential is most suitable when one wants to know the field in a region around the center of the solenoid, where the experiments are usually performed. Of special interest is the spatial homogeneity of the field. For the calculation one has to solve

$$\Delta\psi = 0 \,, \tag{1.34}$$

where the scalar potential ψ is only defined outside current-carrying conductors. Furthermore, we have assumed here a constant permeability of $\mu = \mu_0$. The region of interest shall be an ellipsoid, so that the condition of a single-connected region is also fulfilled (see Sect. 1.1.4).

Because of the cylindrical symmetry we can make an approach with an infinite series of Legendre polynomials for the scalar potential and write (we use spherical coordinates, ϱ, ϑ and φ; because of cylindrical symmetry no φ-component exists)

$$\psi = \sum_{n=0}^{\infty} \left(a_n \varrho^n + b_n \varrho^{-(n+1)} \right) P_n(\cos \vartheta) \,. \tag{1.35}$$

This is the most general form for ψ. Since the potential has to be finite, one can distinguish two regions: a near-field region, where all coefficients $b_n = 0$, and a far-field region, where all coefficients $a_n = 0$. In the center of the solenoid we are, of course, in the near-field regime. The field at a point P with coordinates ϱ, ϑ and φ is hence

$$B_r = \mu_0 \sum_{n=0}^{\infty} \left[a_n n\, \varrho^{n-1} P_{n-1}(\cos \vartheta) - b_n(n+1)\, \varrho^{-(n+2)} P_{n+1}(\cos \vartheta) \right] \,, \tag{1.36}$$

$$B_\vartheta = \mu_0 \sum_{n=0}^{\infty} \left[-a_n \sin \vartheta\, \varrho^{n-1} P'_{n-1}(\cos \vartheta) + b_n \sin \vartheta\, \varrho^{-(n+2)} P'_{n+1}(\cos \vartheta) \right] \,. \tag{1.37}$$

Because of the cylindrical symmetry there is no azimuthal component of the field, and B_r and B_ϑ do not depend on the azimuthal position of the point P. With the well-known recurrence relations for the Legendre polynomials a very efficient algorithm for calculating the potential ψ can be achieved, which converges very rapidly.

The coefficients a_n and b_n have to be found from the boundary conditions. In the case of a solenoid the field along the z-axis is available in the form of an analytical expression. This expression is then expanded into a Taylor series and by comparison the coefficients a_n and b_n can be determined.

One should emphasize that the coefficients a_n and b_n depend only on the coil geometry and the distribution of the current density, and not on the coordinates of the point P. The field sources and the field coordinates are separated. Expressions for the field at a point P are continuous scalar functions of the coordinates of P, the information about the coil is hidden in the coefficients a_n and b_n. This implies that $\boldsymbol{B}$ can be written as scalar function of the field coordinates with constant coefficients depending only on the source coordinates. Furthermore, $\boldsymbol{B}$ can be directly integrated or differentiated. The method shown here is also known as zonal harmonic analysis [25–29]. We conclude that zonal harmonic analysis is the ideal tool if only the field outside the winding volume is of interest.

Example. As an example we demonstrate the above procedure for the field around the center of a solenoid with constant current density. In the region around the center there is no current flowing, so we can introduce a scalar magnetic potential. Expressing the scalar potential as a series in Legendre polynomials as in (1.35) we can determine the coefficients by a comparison with the field along the z-axis, which can be found in analytical form. For the field along the z-axis we find (cylindrical coordinates)

$$B_z(0,z) = \frac{1}{2}\mu_0 j \left[(z+b) \ln \frac{a_2 + \sqrt{a_2^2 + (z+b)^2}}{a_1 + \sqrt{a_1^2 + (z+b)^2}} \right.$$
$$\left. -(z-b) \ln \frac{a_2 + \sqrt{a_2^2 + (z-b)^2}}{a_1 + \sqrt{a_1^2 + (z-b)^2}} \right]. \tag{1.38}$$

Now we make a Taylor series around the point $z = 0$:

$$B_z(0,z) = \sum_{n=0}^{\infty} \frac{1}{n!} \partial_n B_z(0,z) \Big|_{z=0} z^n. \tag{1.39}$$

All partial derivatives can be calculated from (1.38) and finally we can identify the coefficients a_n and b_n from (1.36) and (1.37). The first few coefficients a_n are summarized in Table 1.1; all b_n disappear. With these coefficients we can now calculate the field also on points off the z-axis.

The Magnetic Vector Potential A. The magnetic vector potential A allows us to calculate the magnetic field B everywhere, especially within the coil volume. Though the vector potential can be regarded only as an intermediate quantity for calculating the field, it is however very practical for calculating other quantities, for instance the inductances (see 1.32). The vector potential for the solenoid is

$$\boldsymbol{A}(\boldsymbol{r}) = \frac{\mu_0}{4\pi} \int \frac{\boldsymbol{j}(\boldsymbol{r}')}{|\boldsymbol{r} - \boldsymbol{r}'|} \, dV', \tag{1.40}$$

Table 1.1. Zonal harmonic analysis of the central field of a coil with constant current density j_0. The table shows the first few Legendre polynomials and the first few non-zero coefficients a_n. All coefficients b_n disappear: $b_n \equiv 0$

Legendre polynomials
$P_0(x) = 1$
$P_1(x) = x$
$P_2(x) = \frac{1}{2}(3x^2 - 1)$
$P_3(x) = \frac{1}{2}(5x^3 - 3x)$
$P_4(x) = \frac{1}{8}(35x^4 - 30x^2 + 3)$

Coefficients
$a_1 = \frac{j_0 r_0}{2} \, 2\beta \ln \frac{\alpha + \sqrt{\alpha^2 + \beta^2}}{1 + \sqrt{1 + \beta^2}}$
$a_3 = \frac{j_0 r_0}{2} \frac{1}{3\beta} \left[\frac{1}{(1+\beta^2)^{\frac{3}{2}}} - \frac{\alpha^3}{(\alpha^2+\beta^2)^{\frac{3}{2}}} \right]$
$a_5 = \frac{j_0 r_0}{2} \frac{1}{60\beta^3} \left[\frac{(20\beta^4 + 7\beta^2 + 2)}{(1+\beta^2)^{\frac{7}{2}}} - \frac{\alpha^3 (20\beta^4 + 7\beta^2 \alpha^2 + 2\alpha^4)}{(\alpha^2+\beta^2)^{\frac{7}{2}}} \right]$

where the integration is carried out over the coil volume V'.

In a computer program the volume integration is carried out by dividing the coil volume into n cells, wherein a constant current density is assumed. Then the contributions of all cells are summed up [30]. By successively subdividing the cells into smaller ones a converging series of values for the integral is achieved. This iteration process is stopped as soon as two successive results differ by less than a given value ε from each other. It is then said that the integral is accurate within ε. The computational effort for this algorithm is of the order of n^3 (n integration cells in each direction of space) with the additional difficulty that the vector character of $\boldsymbol{j}$ has to be taken into account, which is the main disadvantage of this direct approach. Improvements are possible by an adaptive refinement of the integration volume. This means that the volume is divided into small cells where there are rapid changes in the integrand, and a coarse mesh is sufficient where the integrand is nearly constant.

In the special case of axial symmetry and of some additional features typical in coil design such as constant current density or current density proportional to $1/r$ (Bitter coils) and coils with the shape of solenoids, an analytical treatment of (1.40) is possible.

At first we make use of the axial symmetry, from which it follows that in cylindrical coordinates only the azimuthal component of $\boldsymbol{A}$ and $\boldsymbol{j}$ is not identical to zero. The z-axis is assumed here as the axis of symmetry.

$$A_\varphi(\boldsymbol{r}) = \frac{\mu_0}{4\pi} \int \frac{j(r', z')}{|\boldsymbol{r} - \boldsymbol{r}'|} \, dV' \ . \tag{1.41}$$

Depending on the order in which the integration over φ', r' and z' is carried out, different procedures are found. From a geometric point of view this is equivalent to composing the coil volume from different elementary units. One can distinguish between ring elements, elements with the shape of cylindrical tubes or wedge elements [20, 31–52].

Ring-Shaped Elements. We compose the coil volume out of elements in the shape of rings with infinitesimal cross-section; see Fig. 1.4. In a cylindrical coordinate system the vector potential of such a ring has only a non-vanishing component in the azimuthal direction of the form

$$A_\varphi(\boldsymbol{r}) = \frac{\mu_0}{4\pi} \int_0^{2\pi} d\varphi' \frac{(j \, dr' \, dz') r' \cos \varphi'}{\left[(r - r' \cos \varphi')^2 + (r' \sin \varphi')^2 + (z - z')^2 \right]^{\frac{1}{2}}} \ , \tag{1.42}$$

which can be expressed with elliptic integrals and therefore written as

$$A_\varphi(\boldsymbol{r}) = \frac{\mu_0 I}{k\pi} \sqrt{\frac{r'}{r}} \left[\left(1 - \frac{k^2}{2} \right) K(k) - E(k) \right], \tag{1.43}$$

with

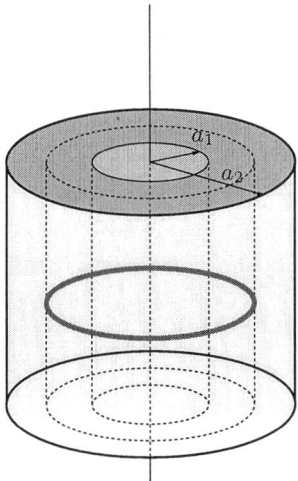

Fig. 1.4. Calculation of the vector potential of a solenoid. The cylindrical coil volume is composed out of many, infinitesimally thin ring elements. Mathematically speaking we integrate first along the azimuthal coordinate in the volume integral (1.41)

$$k = \frac{2\sqrt{r'r}}{\sqrt{(r+r')^2 + (z-z')^2}} \tag{1.44}$$

and E and K as the complete elliptic integrals of the first and second kind:

$$E(k) = \int_0^{\frac{\pi}{2}} d\theta \left[1 - k^2 \sin^2\theta\right]^{\frac{1}{2}}, \tag{1.45}$$

$$K(k) = \int_0^{\frac{\pi}{2}} d\theta \left[1 - k^2 \sin^2\theta\right]^{-\frac{1}{2}}. \tag{1.46}$$

These integrals can be rapidly calculated by the method of the arithmetic-geometric mean (AGM-method, [52–54]). To obtain the vector potential of the whole solenoid, two numerical integrations, one in the radial and one in the axial direction, are necessary.

The advantage of this method is that any arbitrary coil, which means the shape as well as the current density distribution, can be calculated, as long as there exists axial symmetry.

Compared to the following methods, the method with ring-shaped elements is relatively slow, however, and it becomes singular if one tries to calculate the potential of a point very close to the ring element, which is the main drawback of this way of calculating the vector potential of a solenoid.

Elements in the Form of Thin Cylinders. For this method the basic volume element consists of an infinitesimal thin cylinder with the total height

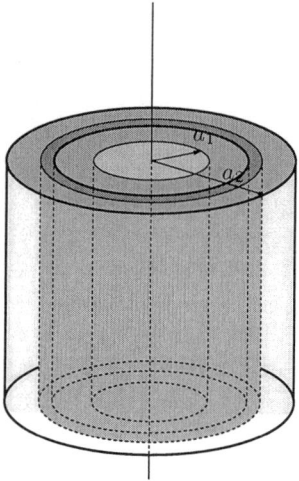

Fig. 1.5. Decomposing a cylindrical coil into a set of infinitesimally thin cylinders. The vector potential of such a thin cylinder can be expressed in terms of elliptic integrals; see (1.47)

of the coil, which means one carries out first the integration along the axial direction and then in the azimuthal direction (see Fig. 1.5). This z-integration can often be done in analytical form, the result again expressed in terms of elliptic integrals of the first, second and the third kind [50]:

$$A_\varphi(\boldsymbol{r}) = \frac{\mu_0 I}{4\pi} \frac{z'-z}{2r} \left[\frac{(z'-z)^2 + 2r'^2 + 2r^2}{a} K(k) \right.$$
$$\left. - a\, E(k) - \frac{(r'-r)^2}{a} \Pi(n,k) \right]\Bigg|_{z'=-b}^{z'=b}, \qquad (1.47)$$

with the abbreviations

$$a^2 = (r'+r)^2 + (z-z')^2, \qquad (1.48)$$

$$k^2 = \frac{4rr'}{a^2}, \qquad (1.49)$$

$$n^2 = \frac{4rr'}{(r+r')^2}, \qquad (1.50)$$

$$\Pi(n,k) = \int_0^{\frac{\pi}{2}} d\theta \, \left[1 + n^2 \sin^2\theta\right]^{-1} \left[1 - k^2 \sin^2\theta\right]^{-\frac{1}{2}}, \qquad (1.51)$$

where I denotes the current through the cylinder. Now only one additional integration in the radial direction has to be carried out to calculate the vector potential of the solenoid, which makes this approach superior to that with the ring elements. The method is applicable as long as one is able to carry out the integration in the axial direction by hand, which holds for a constant current density, for instance. This may not be true for a more complicated form of the current distribution.

Wedge Elements. Here we first integrate along the radial and axial directions. If these integrations can be carried out in an analytical way – two important cases are the coil with constant current density and the Bitter coil with current density proportional to $1/r$ – then a final numerical integration along the azimuthal direction results in the vector potential of the solenoid. The situation is sketched in Fig. 1.6. For a coil with rectangular cross-section and constant current density we find

$$A_\varphi(\mathbf{r}) = \frac{\mu_0 j}{4\pi} \int_0^{2\pi} d\varphi' \Big[-r\sin^2\varphi'\,(z-z')\ln\left(\sqrt{} + r' - r\cos\varphi'\right)$$
$$+ 2r^2 \sin^2\varphi' \cos\varphi' \ln\left(\sqrt{} + z - z'\right)$$
$$+ r^2 \sin\varphi' \left(\sin^2\varphi' - \cos^2\varphi'\right) \arctan\Theta \Big] \Bigg|_{r'=a_1}^{r'=a_2} \Bigg|_{z'=-b}^{z'=b} , \quad (1.52)$$

with the abbreviations

$$\sqrt{} = \sqrt{(r - r'\cos\varphi')^2 + (r'\sin\varphi')^2 + (z - z')^2} , \quad (1.53)$$

$$\Theta = \frac{(z-z')\sqrt{} + (z-z')^2 + (r\sin\varphi')^2}{(r' - r\cos\varphi')r\sin\varphi'} . \quad (1.54)$$

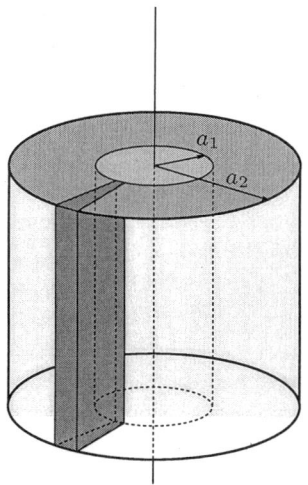

Fig. 1.6. Calculation of the vector potential of a solenoid by decomposing the coil into infinitesimally thin wedge elements. Mathematically speaking the volume integration is first carried out over the radial and the axial directions. For the coil type with constant current density and the Bitter coil these two integrations can be carried out in closed form

16   1. Basics

For a Bitter coil with a current density distribution proportional to $1/r$ we get

$$A_\varphi(r) = \frac{\mu_0 j_0}{4\pi} a_1 \int_0^{2\pi} d\varphi' \left[r \sin^2\varphi' \ln\left(\sqrt{} + z - z'\right) \right.$$

$$\left. - r \sin\varphi' \cos\varphi' \arctan\Theta \right] \Bigg|_{r'=a_1}^{r'=a_2} \Bigg|_{z'=-b}^{z'=b} . \tag{1.55}$$

The final calculation of the vector potential now requires only a 1-dimensional numeric integration along the azimuth, for which very fast algorithms exist. Wedge elements are therefore the preferred method when calculating $\boldsymbol{A}$.

The magnetic field follows by building the rotation of the vector potential. In general this is not the best way to calculate $\boldsymbol{B}$, because in applying the rotation to $\boldsymbol{A}$ one has – on the numeric side – to calculate finite differences of $\boldsymbol{A}$, which is not very accurate. Therefore, a different approach is needed for calculating the $\boldsymbol{B}$-field of a solenoid, namely the Biot-Savart law.

Method with the Law of Biot-Savart. We start with the Biot-Savart law

$$d\boldsymbol{B}(r) = \frac{\mu_0}{4\pi} \frac{\boldsymbol{j}(r') \times (\boldsymbol{r} - \boldsymbol{r}')}{|\boldsymbol{r} - \boldsymbol{r}'|^3} dV' . \tag{1.56}$$

The field at any point P in space is obtained by an integration over the coil volume. The advantage of this method is that any arbitrary configuration with any arbitrary distribution of current densities can be calculated.

$$\boldsymbol{B}(r) = \frac{\mu_0}{4\pi} \int \frac{\boldsymbol{j}(r') \times (\boldsymbol{r} - \boldsymbol{r}')}{|\boldsymbol{r} - \boldsymbol{r}'|^3} dV' . \tag{1.57}$$

As in the previous paragraph a direct volume integration is possible and with arbitrary geometries is the only feasible way to obtain the fields. Again the numerical effort is of the order n^3 with the additional difficulty that three equations for each component of $\boldsymbol{B}$ have to be calculated. Improvements are possible by an adaptive refinement of the integration volume. This means that the volume is divided into small cells where there are rapid changes in the integrand, and a coarse mesh is sufficient where the integrand is nearly constant.

Another improvement is the conversion of the volume integral into a surface integral. In the next three subsections we demonstrate this important technique for a current segment, a solenoid with constant current density and a Bitter coil.

A Current Segment. We consider a volume element throughout which the current density $\boldsymbol{j}$ is constant. The field at a point $\boldsymbol{r}$ produced by this current segment can be found with the help of the segment's vector potential (see Fig. 1.7):

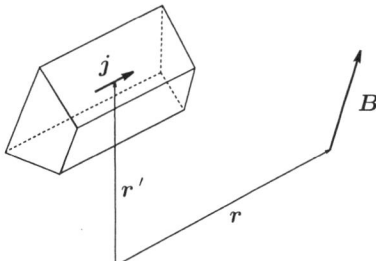

Fig. 1.7. Calculation of the magnetic field $\boldsymbol{B}$ at a point $\boldsymbol{r}$ generated by a current segment. The current density $\boldsymbol{j}$ within the segment's volume shall be constant (magnitude and direction). Points within the segment are denoted by the radius $\boldsymbol{r}'$

$$\boldsymbol{A}(\boldsymbol{r}) = \frac{\mu_0}{4\pi} \int \frac{\boldsymbol{j}}{|\boldsymbol{r}-\boldsymbol{r}'|}\, dV' \,. \tag{1.58}$$

By performing the rotation one gets

$$\boldsymbol{B}(\boldsymbol{r}) = \nabla \times \boldsymbol{A}(\boldsymbol{r}) = \frac{\mu_0}{4\pi} \int \frac{\nabla \times \boldsymbol{j}}{|\boldsymbol{r}-\boldsymbol{r}'|}\, dV' \,. \tag{1.59}$$

Because of the constant current density the unprimed nabla operator can be substituted by a primed nabla operator accompanied by a change in sign.

$$\boldsymbol{B}(\boldsymbol{r}) = -\frac{\mu_0}{4\pi} \int \frac{\nabla' \times \boldsymbol{j}}{|\boldsymbol{r}-\boldsymbol{r}'|}\, dV' \,. \tag{1.60}$$

Using the theorem of Gauss this volume integral can now be transformed into a surface integral:

$$\boldsymbol{B}(\boldsymbol{r}) = \frac{\mu_0}{4\pi} \int \frac{\boldsymbol{j} \times d\boldsymbol{F}'}{|\boldsymbol{r}-\boldsymbol{r}'|} \,. \tag{1.61}$$

The remaining surface integration is numerically of the order n^2, a remarkable improvement.

Generally, for a fast computation of fields or other quantities one should solve the problem under investigation as far as possible by analytical methods first and only then start with numerical methods. Furthermore, the surface integration in (1.61) can be solved so that an analytical expression for $\boldsymbol{B}$ can be given.

Solenoid with Constant Current Density. By using a cylindrical coordinate system the vector potential of a solenoid (see Fig. 1.8) has only a non-vanishing component in the azimuthal direction. For the field at a point $\boldsymbol{r} = (r, \varphi, z)$ one gets

$$\boldsymbol{B}(\boldsymbol{r}) = \frac{\mu_0}{4\pi} \int \frac{\nabla \times \boldsymbol{j}(\boldsymbol{r}')}{|\boldsymbol{r}-\boldsymbol{r}'|}\, dV' \,. \tag{1.62}$$

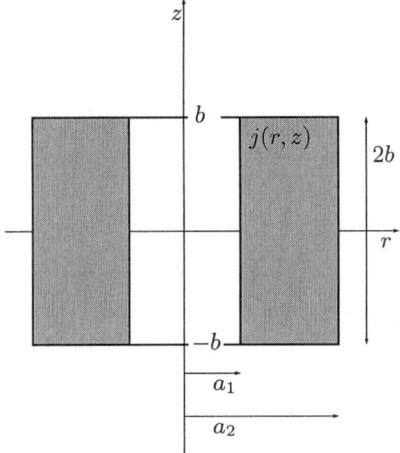

Fig. 1.8. Cross-section of a solenoid with inner radius a_1 and outer radius a_2, extending in the axial direction from $-b$ to b. The current density $j(r, z)$ within the cross-section has only an azimuthal component, and hence the magnetic field of the coil consists only of radial and axial components

Because only the absolute value j_0 of the current density is constant and not $\boldsymbol{j}$, there occurs an additional term when one changes to the primed nabla operator:

$$\nabla \times \frac{\boldsymbol{j}(\boldsymbol{r}')}{|\boldsymbol{r}-\boldsymbol{r}'|} = -\nabla' \times \frac{\boldsymbol{j}(\boldsymbol{r}')}{|\boldsymbol{r}-\boldsymbol{r}'|} + \frac{j_0}{r'}\frac{1}{|\boldsymbol{r}-\boldsymbol{r}'|}\boldsymbol{e}_z \ . \tag{1.63}$$

The additional term gives a contribution in the axial direction $\boldsymbol{e}_z$. The components of the magnetic field can then be calculated to be

$$B_r = \frac{\mu_0 j_0}{2\pi} \int_0^\pi d\varphi' \cos\varphi' \Big[\sqrt{}$$

$$+ r\cos\varphi' \ln\left(\sqrt{} + r' - r\cos\varphi'\right)\Big]\Big|_{r'=a_1}^{r'=a_2}\Big|_{z'=-b}^{z'=b}, \tag{1.64}$$

$$B_\varphi = 0 \ , \tag{1.65}$$

$$B_z = \frac{\mu_0 j_0}{2\pi} \int_0^\pi d\varphi' \Big\{ r\cos\varphi' \ln\left(\sqrt{} + z - z'\right)$$
$$- (z - z') \ln\left(\sqrt{} + r' - r\cos\varphi'\right)$$
$$+ r\sin\varphi' \arctan\theta \Big\}\Big|_{r'=a_1}^{r'=a_2}\Big|_{z'=-b}^{z'=b}, \tag{1.66}$$

with the abbreviations

1.1 Calculation of Magnetic Fields

$$\sqrt{} = \sqrt{(r - r'\cos\varphi')^2 + (r'\sin\varphi')^2 + (z - z')^2} \,, \tag{1.67}$$

$$\theta = \frac{(r' - r\cos\varphi')\sqrt{} + (r' - r\cos\varphi')^2 + (r'\sin\varphi')^2}{(z - z')r\cos\varphi'} \,. \tag{1.68}$$

Bitter Coil. For a Bitter coil with inner radius a_1 and current density $|j| = j_0 \frac{a_1}{r'}$ the change in the nabla operators (see 1.63) is just

$$\nabla \times \frac{j(r')}{|r-r'|} = -\nabla' \times \frac{j(r')}{|r-r'|} \,, \tag{1.69}$$

which results in following fields:

$$B_r = \frac{\mu_0 j_0 a_1}{2\pi} \int_0^\pi d\varphi' \, \cos\varphi' \, \ln\left(\sqrt{} + r' - r\cos\varphi'\right)\Big|_{r'=a_1}^{r'=a_2} \Big|_{z'=-b}^{z'=b} \,, \tag{1.70}$$

$$B_\varphi = 0 \,, \tag{1.71}$$

$$B_z = \frac{\mu_0 j_0 a_1}{2\pi} \int_0^\pi d\varphi' \, \ln\left(\sqrt{} + z - z'\right)\Big|_{r'=a_1}^{r'=a_2} \Big|_{z'=-b}^{z'=b} \,, \tag{1.72}$$

with the abbreviation

$$\sqrt{} = \sqrt{(r - r'\cos\varphi')^2 + (r'\sin\varphi')^2 + (z - z')^2} \,. \tag{1.73}$$

In the case of the solenoid with constant current density and the Bitter coil the originally three-dimensional integration can be reduced to a one-dimensional integration over the azimuth, for which fast numerical algorithms exist.

Inductances. Asides from the calculation of the fields generated by a magnet system one is also interested in its inductance. The inductance of a coil has a great influence on the load characteristic of a pulsed magnet system; for static magnets it is essential for estimating the damping effects on the electrical noise from the power supply. In a multi-coil system the force between any pair of coils is often calculated with the help of their mutual inductance.

As an application of (1.31) we calculate the self-inductance for a cylindrical coil with either constant current density or for a Bitter coil. Let the coils have inner radius a_1, outer radius $a_2 = \alpha a_1$ and height $h = 2\beta a_1$; see Fig. 1.9 for a view of the solenoid in dimensionless coordinates. The self inductance is (we omit the indices)

$$L = \frac{\mu_0}{4\pi} \frac{1}{I^2} \iint \frac{j \cdot j'}{|r - r'|} \, dV \, dV' \,. \tag{1.74}$$

The current through the coil's cross-section is denoted by I and is calculated by an integration of the current density over the cross-section:

$$I = \int j(r,z) \, dr \, dz \,. \tag{1.75}$$

20   1. Basics

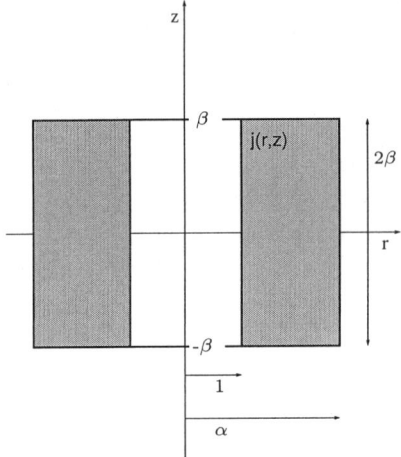

Fig. 1.9. Definition of the dimensionless radii and height of a solenoid, normalized with respect to the inner radius. Hence the inner radius is 1, the outer radius α, and the coil extends in the axial direction from $-\beta$ to β

By measuring all lengths in units of the inner radius a_1 and using a dimensional analysis of (1.74) we can write

$$L = \frac{\mu_0}{4\pi} a_1 \Lambda(\alpha, \beta) \,, \tag{1.76}$$

where we introduce a factor $\Lambda(\alpha, \beta)$, which is called the factor of self-inductance of the solenoid.

So far (1.76) holds only for a coil with one turn. If the coil has N turns of wire or consists of N Bitter plates, we get an additional factor of N^2: If we keep the average current density in the cross-section constant, then the coil generates the same field, independent of the number of turns. The magnetic energy of the field is hence also constant, i.e. $W_m = \frac{1}{2} L I^2 = \text{const}$. Keeping the current density constant means that the current through a wire is proportional to the cross-section of one wire or inversely proportional to the number of turns. For a constant magnetic energy this causes the inductance to be proportional to N^2. We get, therefore, for (1.76) in the general case of N turns or Bitter plates the expression

$$L = N^2 \frac{\mu_0}{4\pi} a_1 \Lambda(\alpha, \beta) \,. \tag{1.77}$$

The factor of self-inductance $\Lambda(\alpha, \beta)$ depends on the shape of the cross-section and the distribution of the currents in the coil. In Figs. 1.10 and 1.11 the factors of self inductance for a wire-wound coil and the Bitter coil are shown, respectively. For the same geometry the coil with constant current density gives higher values for $\Lambda(\alpha, \beta)$ than the Bitter coil.

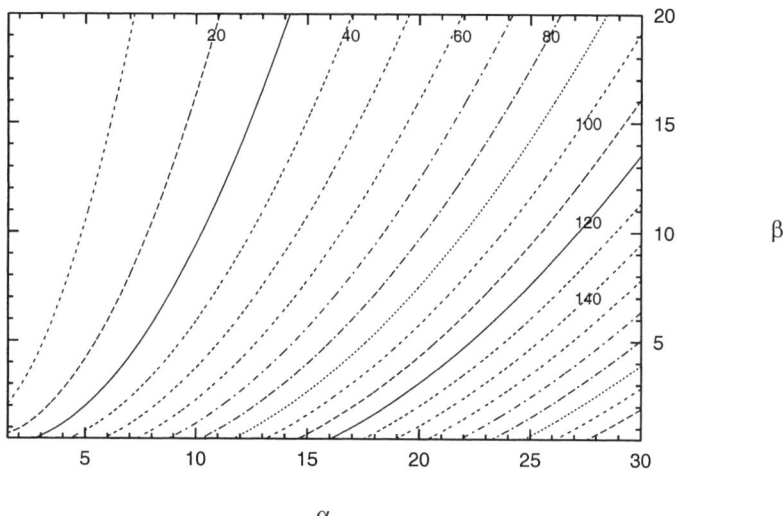

Fig. 1.10. Self-inductance factor $\Lambda(\alpha, \beta)$ for a rectangular coil with constant current density as a function of the shape parameters α and β. For reasons of clarity only a few lines of constant Λ-values are labelled. The self-inductance of a coil with N turns and inner radius a_1 is then $L = N^2 \, \frac{\mu_0}{4\pi} \, a_1 \, \Lambda(\alpha, \beta)$

1.2 Forces and Stresses

There exist two important forces in a coil: the Lorentz forces, caused by the interaction of the current in the conductors with the field of the coil, and thermal forces, caused by a nonuniform temperature distribution in the coil. These forces cause a stress distribution within the coil. In this chapter we consider only stresses induced by the Lorentz forces. We first remember a few basics [55–59], which we then apply to a solenoid.

1.2.1 Elasticity

The theory of elasticity describes the response of a deformable body to the forces acting on it. Under the action of external forces the body gets deformed until it reaches an equilibrium, where the external forces are compensated by internal (molecular) forces within the body. The work done by the external forces during the deformation is completely or partially transformed into potential energy of strain. Upon removal of the external forces the body will return partly or completely to its original shape. In the latter case we speak of an elastic deformation, while the former case is called plastic deformation. The onset of this plastic deformation is a material constant and is called

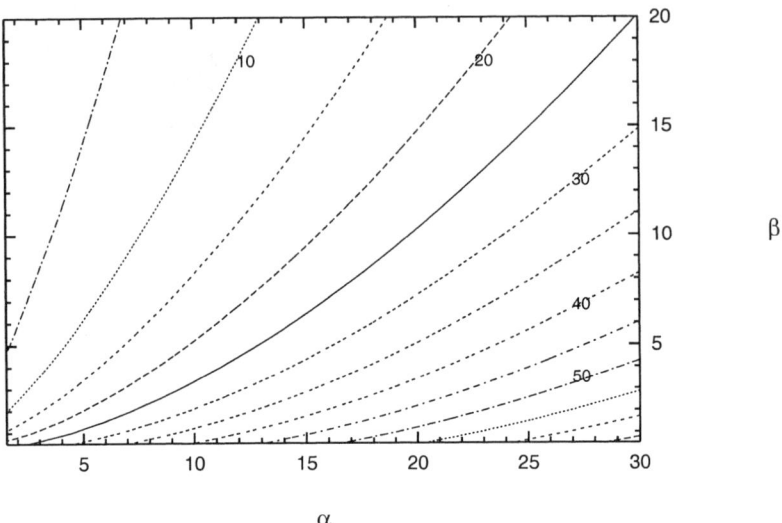

Fig. 1.11. Self-inductance factor $\Lambda(\alpha, \beta)$ for an ideal Bitter coil with current density $\sim 1/r$ as a function of the shape parameters α and β. Only some lines of constant Λ-values are labelled. The inductance of the coil follows $L = N^2 \frac{\mu_0}{4\pi} a_1 \Lambda(\alpha, \beta)$, where a_1 is the inner radius and N the number of Bitter plates

the yield strength. The detection and measurement of this limit presents an experimental problem since it depends on the precision of the measurement of small deformations. To obtain an objective measure the yield strength is usually defined to be the necessary stress for achieving a strain of 0.2%.

1.2.2 Stress and Strain

For many structural materials a linear relationship between the external force and the deformation of the body has been found (Hooke's law). Depending on the material this linear relationship holds only within certain limits of forces and deformations. Hooke found for the linear elongation of a long bar (see Fig. 1.12) that

$$\sigma = E\,\varepsilon. \tag{1.78}$$

Here σ, called the tensile stress, is defined as the pulling force per cross-section of the bar. The tensile strain is denoted by ε and is defined as the relative elongation of the bar. The material property E is called Young's modulus. The units of stress and Young's modulus are Pa, the strain is a dimensionless number.

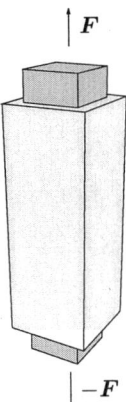

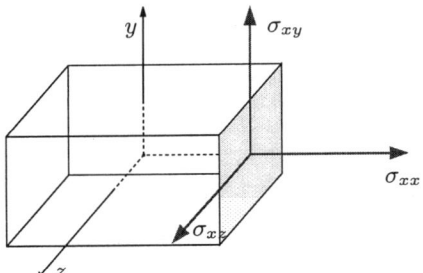

Fig. 1.12. Deformation of a long bar under a force pulling on both ends. The length increases and the cross-section decreases

Fig. 1.13. Normal and tangential stress components at the positive x-surface of a cuboid. The first index in σ_{ik} denotes the direction of the normal vector of the surface, the second one the direction of the force

Aside from an elongation the bar undergoes also a contraction of its cross-section A_q. The contraction of the cross-section, denoted as $-\varepsilon'$ in relative units, leads to a second material constant, the so-called Poisson ratio ν, which is defined as the ratio of the contraction of the cross-section and the elongation of the rod's length:

$$\nu = \frac{-\varepsilon'}{\varepsilon}. \tag{1.79}$$

For a few representative materials the Young's modulus and Poisson ratio are given in Table 1.2.

For a general description of deformations in a three-dimensional body the stress and strain become tensors. In Fig. 1.13 the situation is sketched for the three stress components σ_{xx}, σ_{xy} and σ_{xz}. Any force acting on the gray surface in Fig. 1.13 can be split into three forces in the x, y, and z-directions. These forces divided by the area of the gray surface define the stress components. Similar expressions can be defined for the other surfaces of the cuboid in Fig. 1.13. The components σ_{ii} are called normal stresses and

Table 1.2. Young's modulus E and Poisson ratio ν for a few selected materials

Material	E [GPa]	ν
Aluminum 6061	73	0.33
Copper	117	0.34
Steel 304	193	0.29
Sapphire Al_2O_3	335	0.25

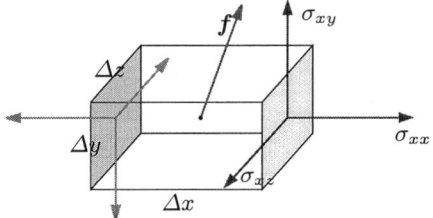

Fig. 1.14. A cuboid with volume $\Delta V = \Delta x\,\Delta y\,\Delta z$ with stresses acting on its surface and under a volume force f. The condition of equilibrium requires that, for instance in the x-direction, the sum of the volume forces and the stresses on the positive and negative x-surfaces (gray areas) cancel each other

the terms σ_{ik} ($i \neq k$) are called tangential stresses. The set of all σ_{ik} form the so-called stress tensor $\boldsymbol{\sigma}$, a symmetric tensor of rank two.

The strain tensor $\boldsymbol{\varepsilon}$ has similar properties. The diagonal elements of ε_{ii} represent the relative changes in the length along the corresponding axes, i.e. elongation or compression in the x, y or z-direction. The three off-diagonal elements ε_{ik} ($i \neq k$) are the shear-components.

1.2.3 Equilibrium of Forces

We consider a small volume element $\Delta V = \Delta x\,\Delta y\,\Delta z$ as shown in Fig. 1.14. The cuboid is subjected to stresses acting on its surface and we allow also an external force acting on the volume, represented by its force density f. The equilibrium of forces in the x-direction yields:

$$\frac{\partial \sigma_{xx}}{\partial x} + \frac{\partial \sigma_{yx}}{\partial y} + \frac{\partial \sigma_{zx}}{\partial z} + f_x = 0 . \tag{1.80}$$

Similar equations are found for the y- and z-directions; in shorter notation the equilibrium of forces is written as

$$\nabla \boldsymbol{\sigma} + \boldsymbol{f} = 0 . \tag{1.81}$$

Equation (1.81) is not sufficient to calculate the stress tensor $\boldsymbol{\sigma}$ for a given force density f, however, because there are only three equations for the six coefficients of $\boldsymbol{\sigma}$. Three more equations are needed; these are the so-called compatibility relations or stress-strain relations. They connect the stress with the strain in the material. Additionally, boundary conditions on the surface of the body can occur.

1.2.4 Stress-Strain Relations

The derivation of a general stress-strain relation is performed in the framework of the classical elasticity theory [60]. In the most general case a total of 21 elastic constant are necessary. Fortunately, for an isotropic body this number gets reduced to 2 and the stress-strain relation is written as

$$\boldsymbol{\sigma} = 2\mu\,\boldsymbol{\varepsilon} + \lambda\,\mathrm{Tr}(\boldsymbol{\varepsilon}). \tag{1.82}$$

A linear and isotropic material possesses two elastic constants μ and λ, which are called Lamé's coefficients. By comparison one can find expressions for Lamé's coefficients in term of Young's modulus and Poisson ratio:

$$2\mu = \frac{E}{1+\nu}, \tag{1.83}$$

$$\lambda = \frac{\nu E}{(1+\nu)(1-2\nu)}. \tag{1.84}$$

1.2.5 Application to a Solenoid

We apply now the results of the last paragraph to a solenoid [61–74]. Because of the axial symmetry we use cylindrical coordinates. The three components of the divergence of the stress tensor are then

$$(\nabla\boldsymbol{\sigma})_r = \frac{1}{r}\partial_r(r\sigma_{rr}) + \frac{1}{r}\partial_z(r\sigma_{zr}) + \frac{1}{r}\partial_\varphi(r\sigma_{\varphi r}) - \frac{1}{r}\sigma_{\varphi\varphi}, \tag{1.85}$$

$$(\nabla\boldsymbol{\sigma})_\varphi = \frac{1}{r}\partial_r(r\sigma_{r\varphi}) + \frac{1}{r}\partial_z(r\sigma_{z\varphi}) + \partial_\varphi\sigma_{\varphi\varphi} + \frac{1}{r}\sigma_{r\varphi}, \tag{1.86}$$

$$(\nabla\boldsymbol{\sigma})_z = \frac{1}{r}\partial_r(r\sigma_{rz}) + \frac{1}{r}\partial_z(r\sigma_{zz}) + \partial_\varphi\sigma_{\varphi z}. \tag{1.87}$$

The rotational symmetry implies that all derivatives with respect to φ disappear and that there exist no shear components along the azimuthal direction, i.e. $\sigma_{r\varphi} = 0$ and $\sigma_{z\varphi} = 0$. Hence the equilibrium condition is written as

$$\frac{1}{r}\partial_r(r\sigma_{rr}) + \partial_z\,\sigma_{zr} - \frac{1}{r}\sigma_{\varphi\varphi} + f_r = 0, \tag{1.88}$$

$$\frac{1}{r}\partial_r(r\sigma_{rz}) + \partial_z\,\sigma_{zz} + f_z = 0. \tag{1.89}$$

The strongest component of the field in a solenoid lies in the axial direction and produces an outward radial force on the conductors. Towards the end of the coil the field has a strong radial component, so that the force perpendicular to this field has a strong axial component, which tries to compress

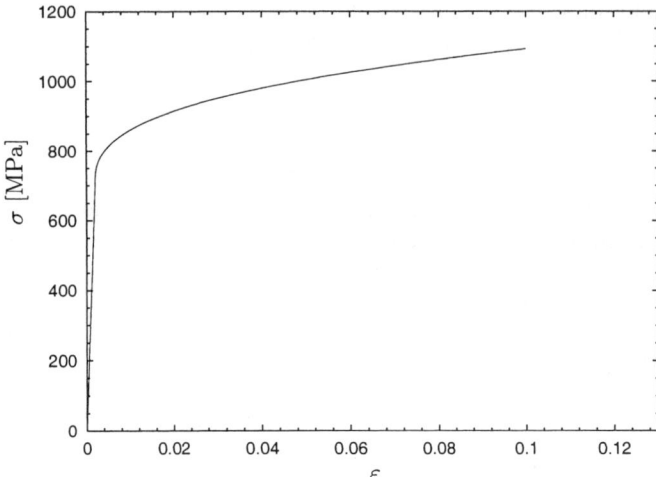

Fig. 1.15. Typical stress-strain or hardening curve for a steel. Up to the elastic limit (about 700 MPa here) the relation between the stress σ and strain ε is linear; above the elastic limit the relation becomes nonlinear; and at a strain of $\varepsilon = 0.1$ the material fails, i.e breaks

the coil. The general stress pattern is therefore circumferential tension and axial compression, compression or tension in the radial direction and various components of shear.

A further simplification occurs in the case when there is no transmission of mechanical forces in the radial direction, as in the case of an ideal polyhelix coil. Then σ_{rr} and σ_{zr} disappear and we get

$$\frac{1}{r}\sigma_{\varphi\varphi} + f_r = 0 , \tag{1.90}$$

$$\partial_z \sigma_{zz} + f_z = 0 . \tag{1.91}$$

1.2.6 Strength Theories

In the simplest models of plasticity a unidirectional force is applied to an isotropic material until plastic (irreversible) strains are induced. After removal of the load plastic (residual) deformations remain. The behavior is described by a stress-strain curve. Phenomenological aspects are derived directly from experiments with the most simply configuration, namely one-dimensional or uniaxial experiments. A typical stress-strain curve is shown in Fig. 1.15. A long as the stress lies below the so-called elastic limit, the relation between stress and strain is linear. As soon as the stress reaches values above the elastic limit there appear irreversible deformations.

The detection of this limit represents an experimental problem since it depends on the precision of the strain-measuring device in use. In the interest of an objective measure a conventional definition of this limit is usually adopted. The conventional elastic limit is the stress which corresponds to the occurrence of a specified amount of permanent strain. For quality control of materials a conventional value of permanent strain equal to 0.2% is commonly used. If the above value of permanent strain is too high, a more refined conventional value of 0.02% is commonly used.

Mechanical properties are normally determined by tests which subject the specimen to simple stress conditions, for instance tensile tests or compression tests. However, the strength of materials under more complicated stress conditions has only been investigated in a few cases. In order to determine suitable allowable stresses various strength theories have been developed.

The generally accepted yield condition for metals is the von Mises condition, which is based on the strain energy of distortion [75–78].

This idea was put forward by J.C. Maxwell in a letter to William Thomson [79]: 'Given the mechanical strain in three directions on an element, when will it give way?' He further writes: 'I have strong reasons for believing that when the strain energy of distortion reaches a certain limit then the element will begin to give way'. For any given stress σ one can calculate the so-called von Mises stress defined as

$$\sigma_{vM} = \sqrt{\frac{3}{2}\text{Tr}(\sigma'^{\,2})} = \sqrt{\frac{3}{2}\left[\text{Tr}(\sigma^2) - \frac{1}{3}\left(\text{Tr}(\sigma)\right)^2\right]}. \tag{1.92}$$

It maps a multi-axial stress state σ onto an uniaxial stress state characterized by σ_{vM}.

According to the von Mises criterion any multi-axial state of stress σ can be represented by the von Mises stress σ_{vM} and then be compared to an uniaxial case. If the von Mises stress is smaller than the yield stress (see Fig. 1.15) then one works in the elastic regime, otherwise in the plastic regime.

1.3 Heating of Pulsed Coils

This chapter deals with the heating process of pulsed coils. The generating current is allowed to flow only for a short time, the heat capacity of the magnet is used as a reservoir. Because of the short heat pulse the heating process can be regarded as being adiabatic. The initial temperature of the coil, the allowed final temperature and the conductor material determine the length of the current pulse. Furthermore, one has to deal with the effects of magnetorestance and of eddy currents.

1.3.1 Heating Process

The current producing the field is only allowed to flow for a short time, during which the temperature of the coil will increase steadily from an initial to a final value. The ohmic heat produced during a time interval dt causes a temperature rise of dT in the conductor:

$$\varrho_{el}(T)\, j_{con}^2(t)\, dt = \varrho_m\, c(T)\, dT \,. \tag{1.93}$$

Here $\varrho_{el}(T)$ describes the electrical resistivity of the conductor, $j_{con}(t)$ the current density in the conductor and ϱ_m the mass density. The specific heat is denoted by $c(T)$. For coils where the current density varies throughout the coil, (1.93) holds only locally and the temperature distribution evolves.

A current pulse lasting from a time t_i up to t_f raises the temperature from an initial value T_i to a final value T_f, which can be found by an indefinite integration of (1.93):

$$\int_{t_i}^{t_f} j_{con}^2(t)\, dt = \int_{T_i}^{T_f} \frac{\varrho_m\, c(T)}{\varrho_{el}(T)}\, dT \,. \tag{1.94}$$

The temperature integral on the right-hand side of (1.94) depends on the conductor material and the integration limits, and the initial and the final temperature. We define

$$\mathcal{F}_{Mat}(T_i, T_f) := \int_{T_i}^{T_f} \frac{\varrho_m\, c(T)}{\varrho_{el}(T)}\, dT \tag{1.95}$$

and call $\mathcal{F}_{Mat}(T_i, T_f)$ the material integral.[3] Data and graphs of the function $\mathcal{F}_{Mat}(T_i, T_f)$ can be found in in Sect. 1.3.3. We write the integral on the left-hand side of (1.94) as

$$\int_{t_i}^{t_f} j_{con}^2(t)\, dt = j_{0,con}^2\, t_{pulse}\, \xi \,, \tag{1.96}$$

with the maximum current density $j_{0,con}$, the length of the current pulse t_{pulse} and a correction factor ξ, which reflects the shape of the current pulse. For a rectangular current pulse we have $\xi = 1$, for a half-period of a sine wave $\xi = \frac{1}{2}$ and for a triangular waveform $\xi = \frac{1}{3}$. For any given material, initial and final temperature, and arbitrary shape of the current pulse we therefore write (1.94) as

$$j_{0,con}^2\, t_{pulse}\, \xi = \mathcal{F}_{Mat}(T_i, T_f), \tag{1.97}$$

and since the generated field B is proportional to $j_{0,con}$ we find

$$B^2\, t_{pulse}\, \xi \sim \mathcal{F}_{Mat}(T_i, T_f) \,. \tag{1.98}$$

---

[3] Sometimes this integral is called the action integral. We do not follow this convention because the action in classical mechanics has the unit J s, and an integration with respect to the temperature would lead to the unit J s K. The material integral has the unit $\frac{A^2 s}{m^4}$, however.

For a given coil the product $B^2 t_{\text{pulse}}$ is constant: one can either generate high fields for a short time or lower fields for longer times. For achieving high fields and long pulse lengths one should seek to make the quantity $\mathcal{F}_{\text{Mat}}(T_i, T_f)$ as big as possible. This means that we must select the proper conductor material and the initial as well as the final temperature.

1.3.2 Material Selection

A straightforward method for finding a good conductor material would be to calculate the material integral $\mathcal{F}_{\text{Mat}}(T_i, T_f)$ for various materials and temperature ranges. A more elegant way can be found with the models of Debye and Grüneisen for the specific heat and the resistivity [80]. This model allows us not only to determine the best material but also indicates the useful lower temperature T_i. According to the Debye model the specific heat per mol at the temperature T is given by

$$c_{\text{mol}} = 9 n k_B \left(\frac{T}{\Theta}\right)^3 \int_0^{\Theta/T} dx \, \frac{x^4 e^x}{(e^x - 1)^2} \,, \tag{1.99}$$

with only one empirical parameter, the Debye temperature Θ. The other symbols are the Boltzmann constant k_B and the mole density n. At low temperatures this leads to $c_{\text{mol}} \sim T^3$; at high temperatures it converges to $c_{\text{mol}} \to 3 n k_B$, the classical result of Dulong and Petit.

The resistivity ϱ_{el} can be separated into a part which is caused by intrinsic defects in the metals and a part describing the contribution of the lattice:

$$\varrho_{\text{el}} = \varrho_{\text{intrinsic}} + \varrho_{\text{lattice}} \,. \tag{1.100}$$

For very low temperatures $T \to 0$ the lattice part disappears and only the intrinsic part $\varrho_{\text{intrinsic}}$ remains, which is also called the residual resistivity. Its value depends solely on the concentration of impurities and defects in the material and can be used to characterize the purity of a metal probe. The lattice part has a temperature dependence of $\varrho_{\text{lattice}} \sim T^5$ for very low temperatures ($T \ll \Theta$) and a high temperature limit of $\varrho_{\text{lattice}} \sim T$. Empirically it was found to be well described by

$$\varrho_{\text{lattice}} = \frac{T}{M \Theta^2} f\left(\frac{T}{\Theta}\right) \,, \tag{1.101}$$

where M is the molar mass of the metal and f is a function with the following properties:

$$f(x) \to 1 \quad \text{for} \quad x \to \infty \,,$$
$$f(x) \sim x^4 \quad \text{for} \quad x \to 0 \,.$$

Expressing the temperature in units of the Debye temperature, $T = \vartheta \cdot \Theta$, and the resistivity in units of its value at the Debye temperature, $\varrho(T) = \varrho(\Theta) r(\vartheta)$, (1.95) becomes

$$\mathcal{F}_{\mathrm{Mat}}(T_{\mathrm{i}},T_{\mathrm{f}}) = \mathcal{F}_{\mathrm{Mat}}(\vartheta_{\mathrm{i}},\vartheta_{\mathrm{f}}) = \frac{\varrho_{\mathrm{m}}\,\Theta}{M\,\varrho_{\mathrm{el}}(\Theta)} \int_{\vartheta_{\mathrm{i}}}^{\vartheta_{\mathrm{f}}} \frac{c_{\mathrm{mol}}(\vartheta)}{r(\vartheta)}\,\mathrm{d}\vartheta\,, \tag{1.102}$$

with a material-dependent factor $\frac{\varrho_{\mathrm{m}}\,\Theta}{M\,\varrho_{\mathrm{el}}(\Theta)}$ and a generalized integral independent of the material properties. Graphs for this integral are given in Fig. 1.16 for the parameter $\vartheta_{\mathrm{f}} = 1$ and a few grades of purity, denoted by the RRR-value (residual resistivity ratio). They show that, depending on the purity, cooling below $\approx 0.05\,\Theta$ for very pure samples gives no further improvement. For not-so-pure samples (RRR = 10) this threshold lies at about $0.2\,\Theta$. For copper with a Debye temperature of 315 K this corresponds to 15 K and 65 K, respectively. Cooling a pulsed coil with liquid helium is therefore inadequate and for non-pure copper, i.e. alloys, cooling with liquid nitrogen is totally sufficient. Since the Debye temperature for metals is typically of the order of a few hundred Kelvin, the last statement can be regarded as generally valid with only a few exceptions.

Data for the factor $\frac{\varrho_{\mathrm{m}}\,\Theta}{M\,\varrho_{\mathrm{el}}(\Theta)}$ are shown in Table 1.3 for various metals. Copper is clearly the best choice, followed by silver, gold, beryllium and aluminum. Cobalt, nickel, zinc, molybdenum, tungsten and iridium comprise the next best group of metals; all other elements are inferior.

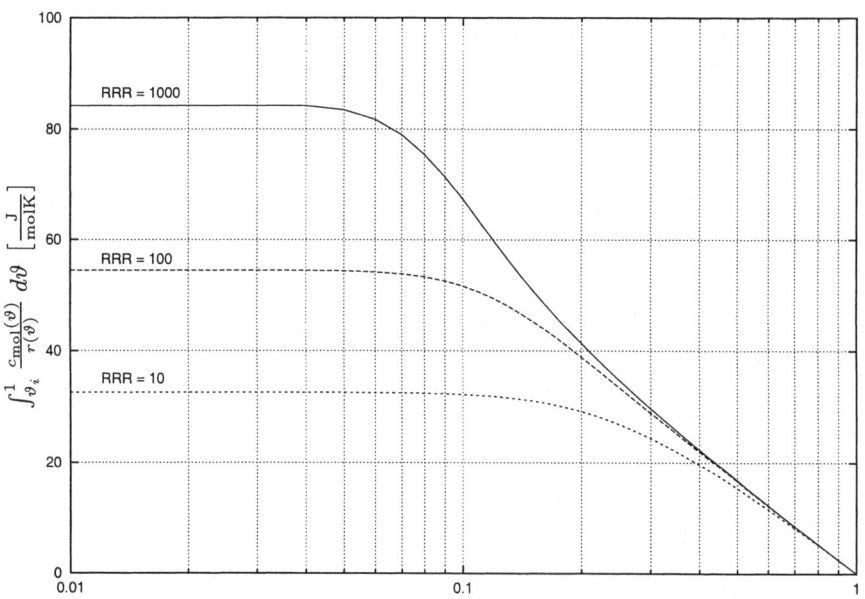

Fig. 1.16. Graph of the integral $\int_{\vartheta_{\mathrm{i}}}^{1} \frac{c_{\mathrm{mol}}(\vartheta)}{r(\vartheta)}\,\mathrm{d}\vartheta$ (see (1.102)) for different purities as a function of the starting temperature ϑ_{i}. The temperatures ϑ and ϑ_{i} are given in units of the Debye temperature Θ. The purities are characterized by their residual resistivity ratio, the RRR-value

Regarding only thermal effects the best material for a pulsed magnet would be copper, cooled down to the temperature of liquid nitrogen (77 K). A cheap and easy-to-realize improvement would be to pump on the liquid nitrogen, thereby reaching about 65 K.

Heat-accumulating capacity is only one limiting factor for a pulsed coil, however; the other is the Lorentz forces. From the latter point of view pure metals are therefore not the best materials, because of their low mechanical strength. One way to increase the strength of a material is to form an alloy. Fortunately the alloying of small amounts – a few percent – of other elements with copper improves its strength with some minor loss in electrical conductivity, so that the superior qualities of copper as a conductor can at least partly be conserved. In amounts of 1%, for example, cadmium increases the strength by 50% with a loss in conductivity of 15%. Other elements for making high-strength copper alloys are beryllium, nickel, chromium, zirconium and silver.

1.3.3 Data for Conductors

For a few representative materials the function $\mathcal{F}_{\mathrm{Mat}}(T_\mathrm{i}, T_\mathrm{f})$ is shown in Fig. 1.17. A few values are shown in Table 1.4. The initial temperature was chosen to be $T_\mathrm{i} = 77\,\mathrm{K}$, and the maximal final temperature to be $T_\mathrm{f} = 400\,\mathrm{K}$. The materials were copper of the grade OFHC, pure aluminum (purity 99.944) and two copper-beryllium alloys, C17510 and C17200. Also shown is the steel AISI 316 and the maragin steel AerMet 100. The data are taken from [95–99].

The values of the function $\mathcal{F}_{\mathrm{Mat}}$ are greatest for copper. Nearly half as good are aluminum and the copper-beryllium alloy C17510. The next best

Table 1.3. Debye temperature Θ and the factor $\frac{\varrho_\mathrm{m}\,\Theta}{M\,\varrho_{\mathrm{el}}(\Theta)}$ for a selection of metals

Metal	Θ [K]	$\frac{\varrho_\mathrm{m}\,\Theta}{M\,\varrho_{\mathrm{el}}(\Theta)}$ [$10^{14}\,\frac{\mathrm{K\,mol}}{\Omega\,\mathrm{m}^4}$]
Be	1000	11.4
Mg	318	4.9
Al	394	10.1
Fe	420	3.5
Co	385	6.5
Ni	375	5.1
Cu	315	24.3
Zn	234	5.5
Mo	380	5.3
Ag	215	18.4
W	310	5.6
Ir	430	6.3
Au	170	13.9

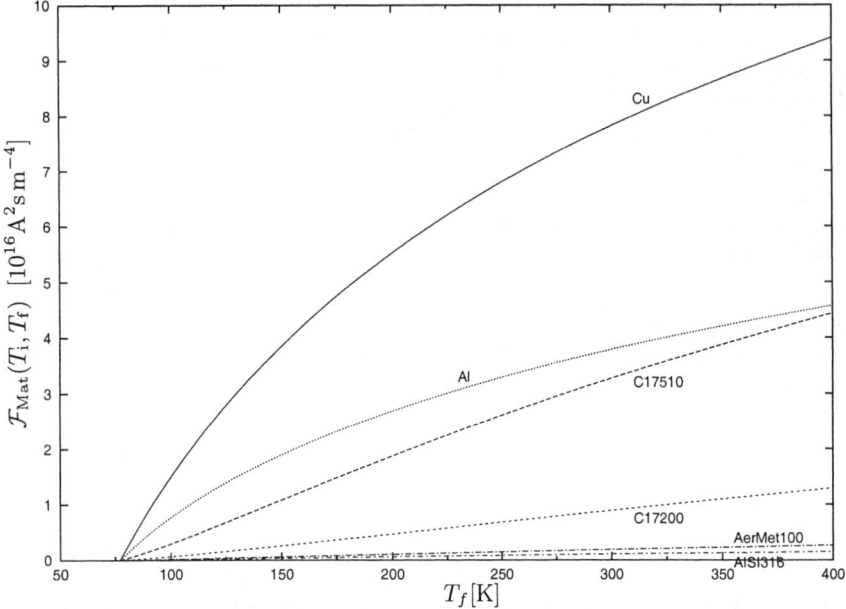

Fig. 1.17. Temperature dependence of the function $\mathcal{F}_{\mathrm{Mat}}(T_\mathrm{i}, T_\mathrm{f})$ from (1.95) for copper, aluminum and two copper-beryllium alloys, C17510 and C17200. The starting temperature is $T_\mathrm{i} = 77\,\mathrm{K}$, and the final temperature T_f varies from 77 K to 400 K. Also shown are the data for the maragin steel AerMet 100 and the steel AISI 316

Table 1.4. $\mathcal{F}_{\mathrm{Mat}}(T_\mathrm{i}, T_\mathrm{f})$ for some materials and temperature ranges

Material	$\mathcal{F}_{\mathrm{Mat}}(T_\mathrm{i}, T_\mathrm{f})$	$\left[10^{16}\,\frac{\mathrm{A}^2\mathrm{s}}{\mathrm{m}^4}\right]$
	77 K → 400 K	77 K → 700 K
Cu	9.42	
Al	4.58	
C17510	4.45	
C17200	1.30	
AerMet 100	0.27	0.46
AISI 316	0.15	0.26

candidate is C17200. The lowest values are for the two steels, with Aermet 100 being slightly better than AISI 316. The case of the steels, especially the maragin steel AerMet 100, is not as bad, however, because they allow a far greater temperature range, for instance 77 → 700 K, and have much higher strength than pure copper or the copper-beryllium alloys. A steel coil would generate the highest fields at relatively short pulse lengths.

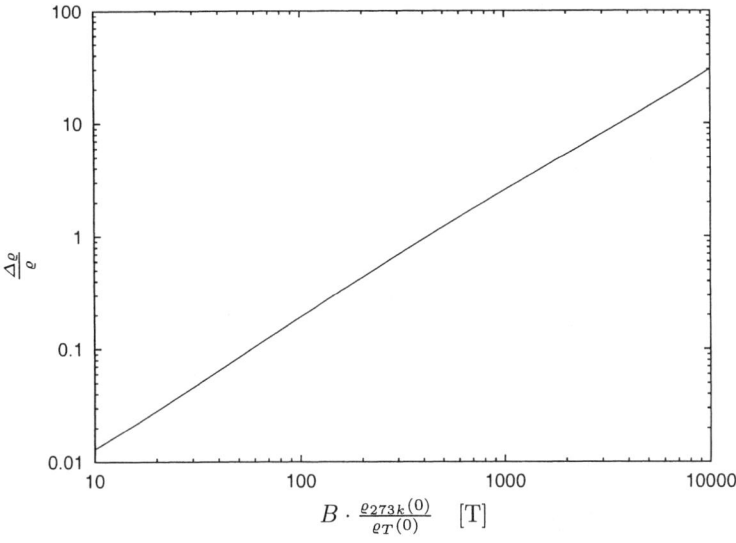

Fig. 1.18. Kohler plot for copper. Shown is the fractional change in resistivity $\Delta\varrho/\varrho$ as a function of the (transverse) magnetic field, divided by the normalized resistivity, $B \cdot \frac{\varrho_{273\mathrm{K}}(0)}{\varrho_T(0)}$. Here $\varrho_{273\mathrm{K}}(0)$ and $\varrho_T(0)$ denote the resistivity at the temperature 273 K and the temperature T at zero field. Normalization of the resistivity permits us to use the Kohler plot also for various degrees of impurities of a pure copper, i.e. for alloys of copper

1.3.4 Magnetoresistance

When exposed to a magnetic field the resistance of a wire increases; this is called magnetoresistance. The effect depends on the orientation of the wire relative to the field as well as the material of the wire. One can discern between transverse and longitudinal magnetoresistance; the latter is generally smaller. Since we deal in this book always with coils of cylindrical symmetry, we need only consider the transverse magnetoresistance.

Generally speaking, due to the magnetic field the path of the conducting electrons experience a deflection and the length of the electron path increases. This leads to an increased scattering rate, which then shows up as an increased resistance.

Magnetoresistance data are usually shown in a so-called Kohler plot, which relates the fractional change in resistivity $\Delta\varrho/\varrho$ to the (transverse) magnetic field divided by the normalized resistivity, $B \cdot \frac{\varrho_{273\mathrm{K}}(0)}{\varrho_T(0)}$. Here $\varrho_{273\mathrm{K}}(0)$ and $\varrho_T(0)$ denote the resistivity at the temperature 273 K and the material temperature T at zero field. The normalized resistivity allows us to use the Kohler plot also for various degrees of impurities of a pure metal. Deviations from the Kohler rule are seen when the wire becomes too thin ('size effect', thickness < 1 mm) or if there are magnetic impurities.

Figure 1.18 shows the average Kohler plot for copper, which one can also use – as just mentioned – for a variety of copper alloys [81]. The increased electrical resistivity due to magnetoresistance has an influence on the heating of pulsed coils, as can be seen from (1.95). The material integral $\mathcal{F}_{\mathrm{Mat}}(T_\mathrm{i}, T_\mathrm{f})$ becomes smaller in the presence of a magnetic field, which in turn causes the possible pulse length t_{pulse} to decrease (see (1.97)), if we assume the current density j_0 and pulse shape factor ξ to remain constant.

The effect of magnetoresistance on a pulsed coil is a change of the temperature distribution in the coil and of the pulse shape. Due to the higher resistance the maximum field is also reduced, as long as one does not compensate for the increased resistance. A simulation is performed in Sect. 4.1.3.

2. Analytical Calculations

This chapter deals with analytical calculations of coils. Quantities of interest are the magnetic field in the center and also within the coil's cross-section, so that the Lorentz forces can be calculated. This leads to the determination of the mechanical stresses. If the distribution of the field is known, then also the magnetic energy and the self-inductance of the coil can be calculated. With the knowledge of the inductance, the resistance (temperature dependent) and the field factor, i.e. the relation of the generated field in the center versus the driving current, the behavior of the coil attached to an energy source like a capacitor bank can be simulated.

Such an analytical treatment would allow for an easy comparison of various magnet designs. Unfortunately, not too much can be calculated the analytical way; for a coil with cylindrical symmetry it is only the field along the symmetry axis. A comparison of the various magnet designs would therefore require us to resort to numerical methods, and an elaborate analysis would then be at least time consuming and mountains of numerical results might hide the underlying basics.

Because of the advantages of an analytical treatment we make some simplifications, so that the field etc. can be determined in a closed form. Of course, this introduces some systematic errors, which are not too big, however, as numerical simulations in Chap. 3 will show.

In Sect. 2.1 we derive the so-called Fabry formula, which gives a relation between the field at the center and the resistive power in the coil. We then apply the Fabry formula to several 'classical' coil types, for instance the wire-wound coil (constant current density), the Bitter coil (current density $\sim 1/r$) and the Gaume and Kelvin current distributions.

For a calculation of the mechanical stresses a further simplification is necessary, which is used in Sects. 2.2 and 2.3. The starting point is a real coil as in Sect. 2.1; then we treat the coil as extended to infinity in the axial direction. This trick allows us to calculate in an analytical way the field and the mechanical stresses in the coil. This method neglects the radial components of the field and the axial components of the stress. It concentrates on the dominant components like the axial field component and the hoop stress.

The general formalism will be developed in Sect. 2.2, which then deals with a special type of coil where there is no transmission of mechanical forces

36   2. Analytical Calculations

in the radial direction. This is called the model of the 'free-standing wire', assuming each wire sustains all Lorentz forces by itself. The 'free-standing wire' model is approximately realized in the polyhelix coil technique, where numerous single-layer coils with different diameters are nested one into each other and where there is no mechanical contact between the different layers.

A comparison of three coil types, the coil with constant current density, the Bitter coil and a stress-optimized coil, is made and finally a few examples are given.

Section 2.3 allows for the transmission of mechanical forces in the radial direction. This model is applied to coils with constant current density as well as to Bitter coils (current density $\sim 1/r$). Additionally we investigate the influence of an outer reinforcement on the stresses in the coil. The chapter closes with a summary of the results for the various coil designs.

2.1 Classical Solenoids

Here various classical solenoids with different distributions of the current density are presented [82–88]. The dimensions of a general cylindrical magnet are shown in Fig. 2.1. Quantities which can be calculated by analytical means are, for instance, the magnetic field on the axis of the solenoid, the resistance or the dissipated ohmic power. The field at points off the symmetry axis or the magnetic energy of the solenoid cannot generally be given in a closed form; one has either make some simplifying assumptions, as will be outlined later, or has to resort to numerical methods.

We will also derive a relation between the dissipated ohmic power and the generated field at the solenoid's center, the classical Fabry formula. The shape of the coil and the type of the current density distribution are represented by a single number, the Fabry factor. The central field is then a function of the Fabry factor, the ohmic power, the resistivity of the conductor in use, the inner radius and the filling factor of the coil. The Fabry formula allows us to compare different coils in terms of field versus power.

Instead of the dissipated ohmic power one can also use the magnetic energy in the coil and construct a second 'Fabry formula', which is a relation between the central field and the magnetic energy in the coil. Again, the shape and the type of the current density distribution are characterized by a (second) Fabry factor. The second 'Fabry formula' is a relation between this factor, the central field, the magnetic energy and the inner radius of the coil. This second Fabry formula is more convenient for superconducting coils and pulsed coils. Again, it serves for comparing different coils in terms of field versus magnetic energy. A calculation of the magnetic energy of a coil can be difficult, since there are no analytical expressions and one has to resort to numerical methods. The experimental determination of the magnetic energy in a coil is easily performed, however, by measuring the inductance of the coil and the current flowing through it.

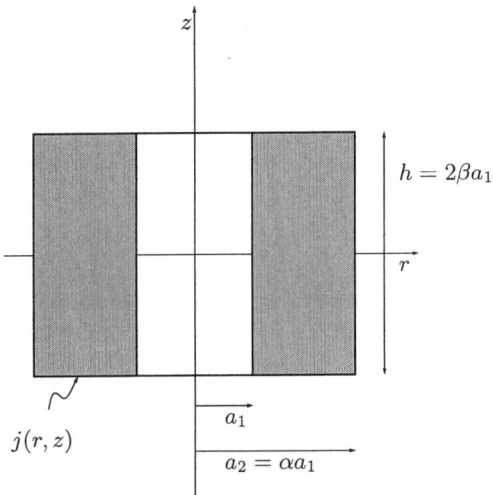

Fig. 2.1. General solenoid with rectangular cross-section and inner radius a_1. The outer radius and height are defined as $a_2 = \alpha\, a_1$ and $h = 2\beta\, a_1$, respectively. In the cross-section there flows a current with density $j(r,z)$; there exists only an azimuthal component of the current density

The efficiency of static coils, given by the Fabry formula, depends on the resistivity of the conductor, on the cross-section of the coil and the distribution of the current density therein. For a coil with rectangular cross-section the Fabry factor is rather low; it can be increased with more sophisticated distributions of the current density, as is done for instance with the Bitter coil or the Gaume coil. The Fabry factor reaches a maximum value for the so-called Kelvin distribution. The maximum possible efficiency is achieved by an infinitely large magnet with the Kelvin distribution and a probe volume with the shape of a sphere.

2.1.1 A Ring-Shaped Current Element

In a general solenoid, as for instance a wire-wound coil, there is a fine structure of the current density throughout its cross-section, namely the conducting wire and non-conducting parts like insulation and glass fiber/epoxy reinforcements. Usually, for the calculation of the magnetic field this finer structure is neglected and an average current density in the cross-section is assumed. For this purpose we separate the current density $j(r,z)$ of the solenoid into the general form

$$j(r,z) = \lambda\, j_0\, f(r,z) \,, \tag{2.1}$$

where j_0 is the maximum current density in the conductor, λ is the filling factor and $f(r,z)$ describes the distribution of the current density.

38   2. Analytical Calculations

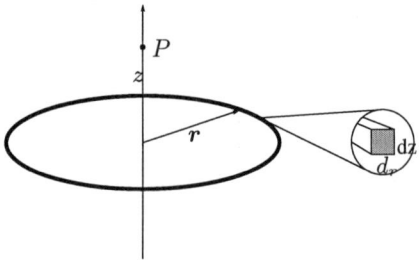

Fig. 2.2. Geometry for calculating the field of a ring-shaped current element with radius r, cross-section $dr\,dz$ and a current density of $j(r,z) = \lambda j_0 f(r,z)$ at a point P on the axis

The separation into a filling factor λ and a current distribution function $f(r,z)$ is mainly for practical reasons, because λ could also be incorporated into $f(r,z)$. For a wire-wound coil with total cross-section A_T, consisting of N turns of wire with cross-section A_W and current I, one has a filling factor of $\lambda = \frac{NA_W}{A_T}$ and a current distribution function of $f = 1$, with $j_0 = \frac{I}{A_W}$ as the current density in the wire. Both λ and j_0 are easily determined. For a general coil one would define the filling factor by

$$\lambda = \frac{\iint_{A_C} f(r,z)\,dr\,dz}{\iint_{A_T} f(r,z)\,dr\,dz}, \tag{2.2}$$

where $f(r,z)$ is the current distribution function, A_T is the total cross-section of the coil and A_C is the area with conductor. j_0 is the maximum current density in the conductor at a certain point. At this point the current distribution function is defined to have the value $f = 1$.

By extraction of λ one can treat the current distribution function $f(r,z)$ as defined over the whole cross-section A_T of the magnet, thereby making the calculations more elegant. The fine structure of conducting and insulating regions in the total cross-section is hidden in the filling factor λ.

Because of the superposition principle the solenoid can be decomposed into a set of infinitesimal thin current rings. The field produced by a ring with radius r at a point P on the axis at a distance z above the ring plane (see Fig. 2.2) may be found by integration of the Biot-Savart law. An infinitesimal element ds of the ring produces a field of

$$\delta \boldsymbol{B} = \frac{\mu_0}{4\pi}\,(\lambda j_0\,dr\,dz)\,\frac{d\boldsymbol{s} \times \boldsymbol{x}}{x^3} = \frac{\mu_0}{4\pi}\,I\,\frac{d\boldsymbol{s} \times \boldsymbol{x}}{x^3}, \tag{2.3}$$

where $\boldsymbol{x}$ is the distance from the point P to the current element $\boldsymbol{s}$. Integration along the ring gives for the produced field

$$dB(z) = \frac{1}{2}\,\mu_0\,\lambda\,j_0\,dr\,dz\,\frac{r^2 f(r,z)}{(r^2+z^2)^{\frac{3}{2}}}. \tag{2.4}$$

For the generated power in the ring one finds

$$dP = \lambda j_0^2 f^2(r,z) \varrho_0 \, dV , \tag{2.5}$$

and by introducing the power density p we find

$$p(r,z) = \frac{dP}{dV} = \lambda j_0^2 f^2(r,z) \varrho_0 . \tag{2.6}$$

With the help of (2.5) and (2.6) one can substitute j_0 in (2.4) and one gets

$$dB(z) = \frac{\mu_0}{2} \sqrt{\frac{\lambda}{\varrho_0}} \sqrt{p(r,z)} \frac{r^2}{(r^2 + z^2)^{\frac{3}{2}}} \, dr \, dz . \tag{2.7}$$

This means that all current rings with the same cross-section $dr \, dz$ lying on the contour line

$$\sqrt{p(r,z)} \frac{r^2}{(r^2 + z^2)^{\frac{3}{2}}} = \text{const} \tag{2.8}$$

will generate the same field dB at the center. In Fig. 2.3 a few of these contour lines are shown for the special case of constant power density, $p = \text{const}$. Each contour line can be characterized by a parameter η, so that the radius where a contour line hits the radial axis is ηa_1. The associated relation (2.8) now becomes

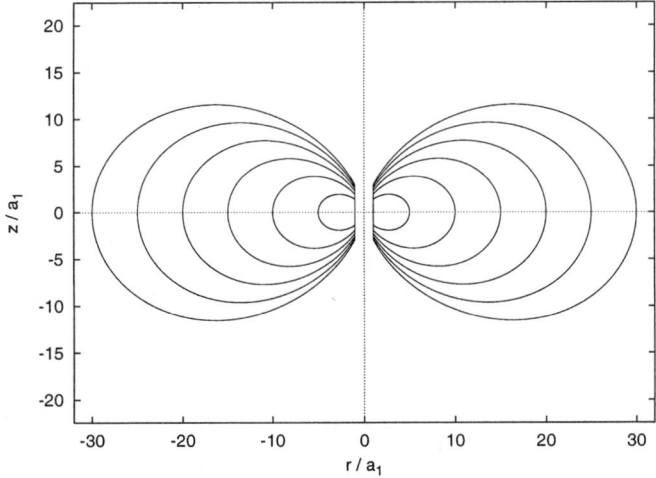

Fig. 2.3. Lines of optimal shape ('apple') for a coil with inner radius a_1 under the assumption of constant power density, $p = \text{const}$. The radial and axial positions of the coil are given in units of a_1. Each contour line can be characterized by its outermost radial position ηa_1, where the contour line hits the radial axis

$$\frac{r^2}{(r^2+z^2)^{\frac{3}{2}}} = \frac{1}{\eta\, a_1}. \tag{2.9}$$

This means that the most efficient current rings in terms of field contribution dB at the center lie on contour lines with low values of η. Current rings close to the center produce the most field.

2.1.2 Calculation of the Fabry Factor

In the relation between the field at the center of a coil and the power producing that field there occurs a factor called the Fabry factor, which is a function of the shape of the coil and the distribution of the currents therein. We note first the general equations for field and power and integrate (2.4) and (2.5). The field at a point with axial coordinate z_p on the solenoid axis is

$$B(z_\mathrm{p}) = \frac{1}{2}\mu_0\,\lambda\,j_0 \iint_A dr\, dz\, \frac{r^2 f(r,z)}{[r^2+(z-z_\mathrm{p})^2]^{\frac{3}{2}}}. \tag{2.10}$$

On the axis there is no radial field component. The field at the center is

$$B_0 = \frac{1}{2}\mu_0\,\lambda\,j_0 \iint_A dr\, dz\, \frac{r^2 f(r,z)}{(r^2+z^2)^{\frac{3}{2}}}. \tag{2.11}$$

For the dissipated ohmic power we find

$$P = 2\pi\,\lambda\, j_0^2\, \varrho_0 \iint_A dr\, dz\, r\, f^2(r,z). \tag{2.12}$$

We now eliminate j_0 and get

$$B_0 = \mu_0 \sqrt{\frac{\lambda P}{\varrho_0}}\, \frac{1}{\sqrt{8\pi}}\, \frac{\iint_A dr\, dz\, \frac{r^2 f(r,z)}{(r^2+z^2)^{\frac{3}{2}}}}{\sqrt{\iint_A dr\, dz\, r\, f^2(r,z)}}. \tag{2.13}$$

If we finally express all lengths in units of a certain unit-length, which is usually chosen to be the radius a_1 of the inner bore of the magnet, that is $r \to u$, $z \to v$ and the cross-section $A \to \tilde{A}$, then we obtain the classical Fabry formula:

$$B_0 = \mu_0 \sqrt{\frac{\lambda P}{\varrho_0 a_1}}\, \underbrace{\frac{1}{\sqrt{8\pi}}\, \frac{\iint_{\tilde{A}} du\, dv\, \frac{u^2 f(u,v)}{(u^2+v^2)^{\frac{3}{2}}}}{\sqrt{\iint_{\tilde{A}} du\, dv\, u\, f^2(u,v)}}}_{=:\ G\ \text{Fabry factor}}. \tag{2.14}$$

This equation states that in order to achieve a high magnetic field one has to have as much power P as possible, the coil should have a high filling factor λ

and consist of a material with low resistivity ϱ_0, and the bore of the magnet should be as small as possible. The Fabry factor G reflects the shape of the cross-section of the magnet and the distribution of the currents therein.

The Fabry formula (2.14) holds for resistive coils in the static case; it is not applicable for superconducting magnets and is inconvenient for pulsed magnets. In the latter cases it is more convenient to express the central field in terms of the stored magnetic energy W_m rather than the power. The current through the total cross-section of the coil is written as

$$I = \lambda j_0 \iint_A dr\,dz\, f(r,z) \,. \tag{2.15}$$

With this current the magnetic energy can be expressed as $W_m = \frac{1}{2} L I^2$. Since the above current is the current flowing through the whole cross-section of the coil, we must identify L as the inductance of a coil with one turn (see 1.76):

$$L = 1^2 \frac{\mu_0}{4\pi} a_1 \Lambda(\alpha,\beta) \,. \tag{2.16}$$

Now the current density in (2.10) can be eliminated and it follows the second Fabry formula:

$$B_0 = \sqrt{\frac{\mu_0 W_m}{a_1^3}} \underbrace{\sqrt{\frac{2\pi}{\Lambda(\alpha,\beta)}} \frac{\iint_{\tilde{A}} du\,dv\, \frac{u^2 f(u,v)}{(u^2+v^2)^{\frac{3}{2}}}}{\iint_{\tilde{A}} du\,dv\, f(u,v)}}_{=:\,\mathcal{G}\ \text{Fabry factor}} \,. \tag{2.17}$$

Here $\mathcal{G}$ denotes a second Fabry factor for superconducting or pulsed coils, which connects the magnetic energy in the coil with the central field. The Fabry factor $\mathcal{G}$ reflects the shape of the coil (integration area) and the type of current density, given by the distribution function $f(u,v)$ and the self-inductance factor $\Lambda(\alpha,\beta)$. High fields are obtained by high values for $\mathcal{G}$ and the magnetic energy W_m, and the inner bore a_1 of the magnet should be as small as possible.

Because of the inductance factor $\Lambda(\alpha,\beta)$, the pulse length (2.17) can no longer be calculated in a closed form. For the rest of this chapter we therefore restrict ourselves to the Fabry formula for the resistive case. We derive expressions for the fields at the center of the coil and the Fabry factor for a few basic coil types. A comparison of the coil types is given at the end of this chapter.

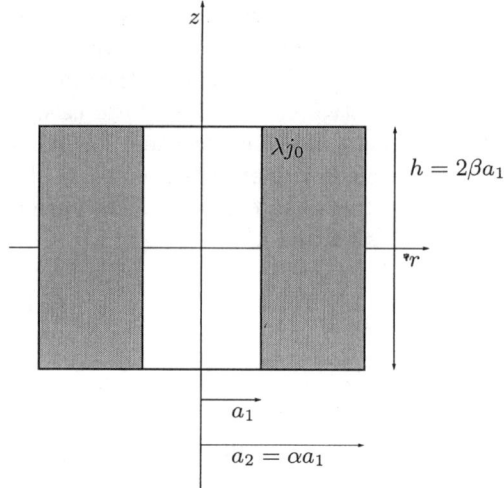

Fig. 2.4. Cylindrical coil with rectangular cross-section and inner radius a_1. The outer radius and height are $a_2 = \alpha\, a_1$ and $h = 2\beta\, a_1$, respectively

2.1.3 Cylindrical Coil with Constant Current Density

This is one of the simplest coils, a wire-wound magnet with cylindrical cross-section. The inner and outer radii are a_1 and a_2, and the height is h. Usually the outer radius and height are expressed in units of the inner radius and we introduce two quantities α and β defined by (see Fig. 2.4)

$$\alpha := \frac{a_2}{a_1}, \quad \beta := \frac{h}{2\,a_1}. \tag{2.18}$$

The constant current density is written as

$$j(r,z) = \lambda\, j_0\, f(r,z) = \lambda\, j_0. \tag{2.19}$$

For the field at the center of the magnet we obtain

$$B_0 = \mu_0 \sqrt{\frac{P\lambda}{\varrho_0 a_1}}\; G(\alpha,\beta), \tag{2.20}$$

where the Fabry factor is defined by

$$G(\alpha,\beta) = \sqrt{\frac{\beta}{2\pi(\alpha^2 - 1)}}\; \ln \frac{\alpha + \sqrt{\alpha^2 + \beta^2}}{1 + \sqrt{1 + \beta^2}}. \tag{2.21}$$

A plot of the Fabry factor is shown in Fig. 2.5; there is a maximum of $G(\alpha,\beta) = 0.1426$ at the point $(\alpha,\beta) = (3.095, 1.862)$.

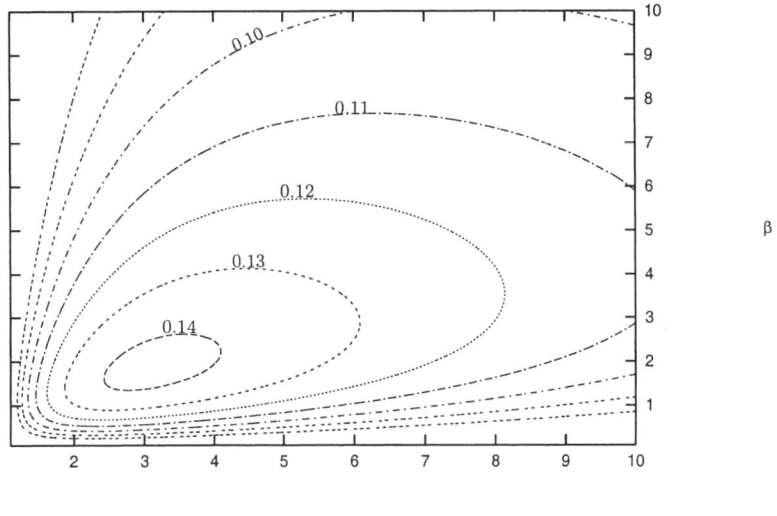

Fig. 2.5. Fabry factor of a rectangular coil with constant current density as a function of the parameters α and β. Contour lines are drawn in steps of 0.01 from 0.08 up to 0.14. The function has a maximum of $G(\alpha, \beta) = 0.1426$ at $(\alpha, \beta) = (3.095, 1.862)$

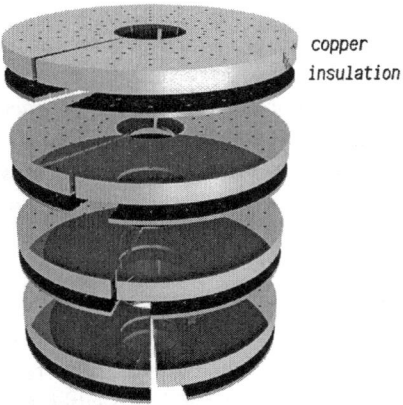

Fig. 2.6. Principal view of a Bitter coil. It is formed by stacking alternating conductor and insulator disks, each with a radial slit. The neighboring conductor parts form a spiral conducting path. The ideal $1/r$ current distribution holds only approximately; it is distorted by the cooling holes

2.1.4 Coil with Current Density $\sim 1/r$

A current density of

$$j(r, z) = \lambda j_0 \frac{a_1}{r} \tag{2.22}$$

44   2. Analytical Calculations

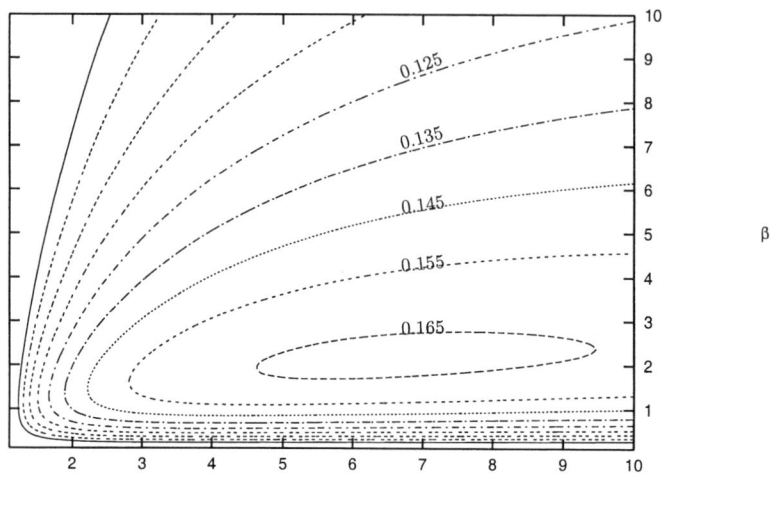

Fig. 2.7. Fabry factor of an ideal Bitter coil with current density $j(r) = \lambda j_0 \frac{a_1}{r}$ as a function of the parameters α and β. There is a maximum at $(\alpha, \beta) = (6.423, 2.146)$ with a value of $G(\alpha, \beta) = 0.1665$. The contour lines run from 0.085 to 0.165 in steps of 0.010; only a few are labelled

is encountered in a so-called Bitter coil [89], which is a stack of alternating conductor and insulator disks with a radial slit and a large number of cooling holes. The neighboring conductor parts are partly contacted with each other to form a spiral conducting path. Figure 2.6 shows the principal design. Because of the cooling holes the $1/r$ current distribution holds only approximately. For the Fabry factor one gets

$$G(\alpha, \beta) = \frac{1}{\sqrt{4\pi\beta \ln \alpha}} \ln \frac{\alpha \left(\beta + \sqrt{1+\beta^2}\right)}{\beta + \sqrt{\alpha^2 + \beta^2}} . \quad (2.23)$$

A graph of $G(\alpha, \beta)$ is shown in Fig. 2.7. Compared to the coil with constant current density a considerable increase in the Fabry factor is possible: the maximum lies at $(\alpha, \beta) = (6.423, 2.146)$ and has a value of $G(\alpha, \beta) = 0.1665$.

2.1.5 The Gaume Current Distribution

Since the step from constant current density to a current density $\sim 1/r$ showed so much improvement, one may also allow a variation in the axial direction. From a practical point of view this can easily be achieved by omitting certain insulating disks in a Bitter coil (see Fig. 2.6), such a coil is called a Gaume coil. We first have to derive the new current distribution and start with a separation of coordinates of the form

2.1 Classical Solenoids

$$j(r,z) = \lambda j_0\, f(r,z) \quad \text{with} \quad f(r,z) = \frac{a_1}{r}\, g(z)\,. \tag{2.24}$$

With this function one has to calculate the magnetic field and the power (2.11 and 2.12), and with the method of Lagrangian parameters the unknown function $g(z)$ can be calculated. The resulting form of the current distribution function $f(r,z)$ is

$$f(r,z) = \frac{\alpha}{\alpha-1}\frac{a_1}{r}\left[\frac{1}{\sqrt{1+(\frac{z}{a_1})^2}} - \frac{1}{\sqrt{\alpha^2+(\frac{z}{a_1})^2}}\right]. \tag{2.25}$$

Now the Fabry factor can be calculated:

$$G(\alpha,\beta) = \frac{1}{\sqrt{4\pi\alpha\ln\alpha}}\sqrt{\alpha\arctan\beta + \arctan\frac{\beta}{\alpha} - 2F\left(\arctan\beta,\sqrt{\frac{\alpha^2-1}{\alpha^2}}\right)}. \tag{2.26}$$

Here $F(\varphi,k)$ denotes Legendre's elliptic integral of the first kind. A graph of the Fabry factor $G(\alpha,\beta)$ can be seen in Fig. 2.8. There is no local maximum anymore, but a global maximum does exist. It has a value of $G = 0.1854$ for $\alpha = 7.762$ and $\beta \to \infty$. Along the line $\alpha = 7.762$ this maximum is very flat; the contour line to $G = 0.185$ can already be seen in the upper right-hand corner in Fig. 2.8, so that for practical purposes a nearly optimal Gaume coil could have $\alpha = \beta = 8$ with an associated Fabry factor of 0.1851.

2.1.6 The Kelvin Current Distribution

We search now for the best current density of all and start with a general current density of the form

$$j(r,z) = \lambda j_0\, f(r,z)\,. \tag{2.27}$$

We follow the same procedure as with the Gaume coil and calculate first the field B and the power P (2.11 and 2.12) and determine by the method of Lagrangian parameters the function $f(r,z)$:

$$\partial_f (B - \eta P) = 0 \tag{2.28}$$

with an arbitrary parameter η yields

$$f(r,z) = \frac{a_1^2\, r}{(r^2+z^2)^{3/2}}\,. \tag{2.29}$$

This is the so-called Kelvin current distribution. For the corresponding Fabry factor we get

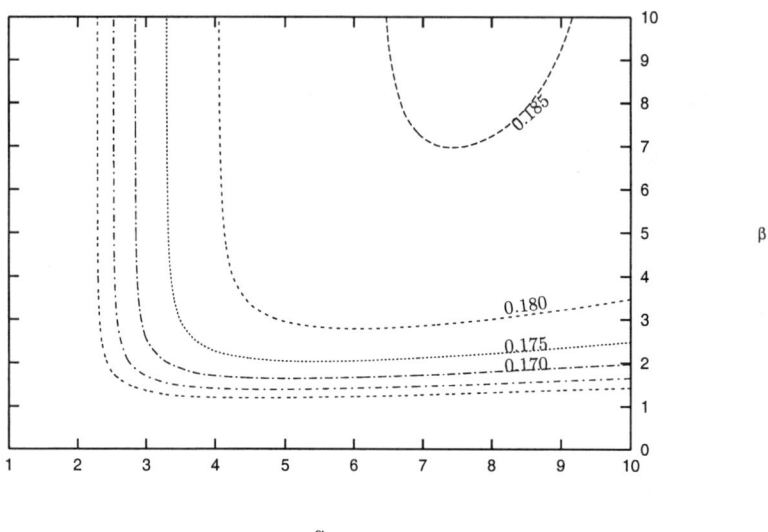

Fig. 2.8. Contours in steps of 0.05 between 0.160 and 0.185 of the Fabry factor of a Gaume coil as a function of the parameters α and β. There is a global maximum of $G(\alpha, \beta) = 0.1854$ for $\alpha = 7.762$ and $\beta \to \infty$. The maximum is very flat in the direction of β, so that nearly optimal Gaume coils can be found within the contour line to $G = 0.185$

$$G(\alpha, \beta) = \frac{1}{\sqrt{32\pi}} \sqrt{3 \left(\arctan \beta - \frac{1}{\alpha} \arctan \frac{\beta}{\alpha} \right) + \beta \left(\frac{1}{1 + \beta^2} - \frac{1}{\alpha^2 + \beta^2} \right)}. \tag{2.30}$$

A graph of the Fabry factor is shown in Fig. 2.9. $G(\alpha, \beta)$ has a global maximum of $G = \sqrt{\frac{3}{64}} \approx 0.2165$ for $\alpha, \beta \to \infty$. As can be seen from Fig. 2.9, it is a very flat maximum; an almost optimal magnet with a finite volume may be defined by the parameter region enclosed by the lines $\alpha = 10$, $\beta = 10$ and the contour line of G with the value 0.20. The technical realization of a current density varying in radial and axial direction may be achieved with the polyhelix technique [90–94], as seen in Fig. 2.10.

2.1.7 The Optimal Magnet with Kelvin Distribution

The magnet in the previous section still had a cylindrical bore of radius a_1, and reached a maximum for $\alpha, \beta \to \infty$. We now allow current to flow also in the bore, except in a probe volume of the form of a sphere with radius a_1 at the center of the magnet. The optimal current distribution is again the Kelvin distribution. The derivation is analogous to that presented in the last chapter, and the Fabry factor is calculated as

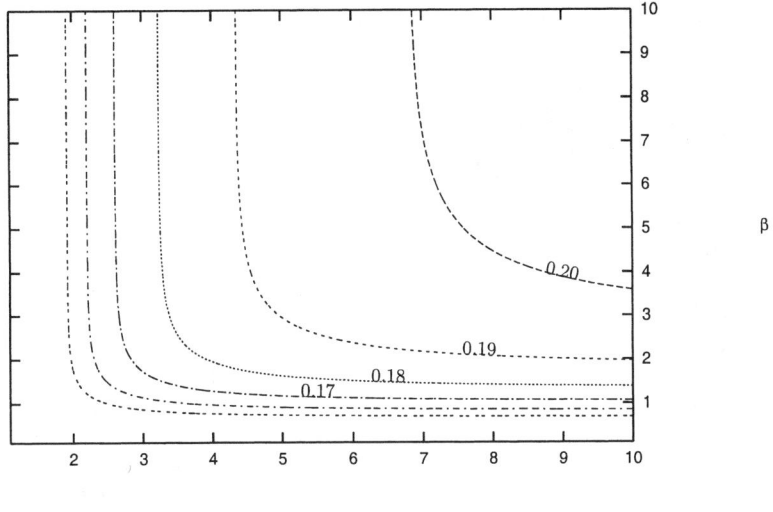

Fig. 2.9. Fabry factor of a coil with the Kelvin current distribution as a function of the parameters α and β. A global maximum of $G(\alpha,\beta) = 0.2165$ occurs for $\alpha,\beta \to \infty$. The maximum is not very sharp; almost optimal configurations with a Fabry factor $G > 0.20$ can be found within the region $\alpha < 10, \beta < 10$. The contour lines are drawn from 0.15 to 0.20 with a step size of 0.01

$$G(\alpha) = \frac{1}{\sqrt{6\pi}} \sqrt{\frac{\alpha-1}{\alpha}} \tag{2.31}$$

for a magnet with the shape of a sphere with outer radius $\alpha\, a_1$. In the limit of $\alpha \to \infty$ we obtain

$$G(\alpha \to \infty) = \frac{1}{\sqrt{6\pi}} \approx 0.2303 \ . \tag{2.32}$$

This is the maximum possible Fabry factor. A coil with this distribution of the current density generates for a given amount of power the maximum possible field.

2.1.8 Coils with Optimal Shape

For a magnet with constant current density ($f(r,z) \equiv 1$) and the optimal shape as given in (2.8), the Fabry factor is only a function of the size of the magnet, which one can characterize by a factor α, defined by expressing the outermost radius of the coil as $a_2 = \alpha\, a_1$. The maximum value is $G = 0.1439$ at $\alpha = 3.309$. The maximum is not very sharp; in the interval $[2.5; 4.5]$ the Fabry factor is still greater than 0.14 (see Fig. 2.11).

Compared to a rectangular coil with constant current density, which has a maximum Fabry factor of 0.1426 for $\alpha = 3.095$ and $\beta = 1.862$, the magnet

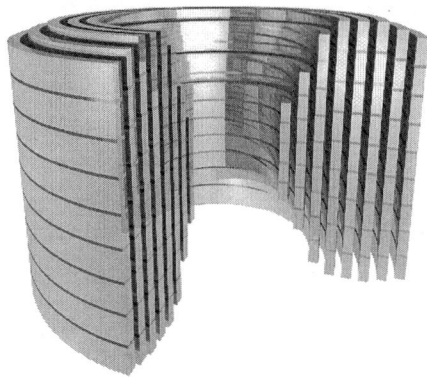

Fig. 2.10. Principle of a polyhelix coil. It consists of several coaxial helices which are supported independently. The electrical connection of the helices is preferably a serial one. Then the same current flows through each winding, and hence the current density is a function of the cross-section of the actual winding. An axial variation of the current density can be realized if the spiral path of a helix is fabricated by cutting a solid tube. This may be done by spark erosion or by cutting with a water jet. Additional parameters to chose from can be the thickness and the height of each tube. Furthermore, one can select different materials for each tube. Polyhelix coils were pioneered by Sir Martin Wood, H.J. Schneider-Muntau and Y. Date [90–94]

with the optimal shape results only in minor improvements. For comparison the Fabry factor for the same coil with the Kelvin current distribution is also shown.

2.1.9 Conclusion

For convenience the results for the cylindrical coils of this chapter are repeated here. For the four investigated coil types – the coil with constant current density, the Bitter coil, the Gaume coil and the Kelvin coil – we list the Fabry factor and additionally expressions for the resistance and the time constant.

The resistance can be found by the equation of the power dissipated in the magnet, $P = RI^2$. By use of (2.12) and expressing the current as an integral of the current density over the cross-section, we find the resistance to be

$$R = N^2 \frac{\varrho_0}{\lambda a_1} 2\pi \frac{\int_1^\alpha du \int_{-\beta}^\beta dv\, u f^2}{\left(\int_1^\alpha du \int_{-\beta}^\beta dv\, f\right)^2} \,. \tag{2.33}$$

The time constant of the coil is defined as $\tau = L/R$. The inductance L can be written in terms of a length, usually chosen to be the inner radius a_1, the number of turns N and a geometry factor Λ reflecting the shape of the cross-section and the current distribution (see (1.76)):

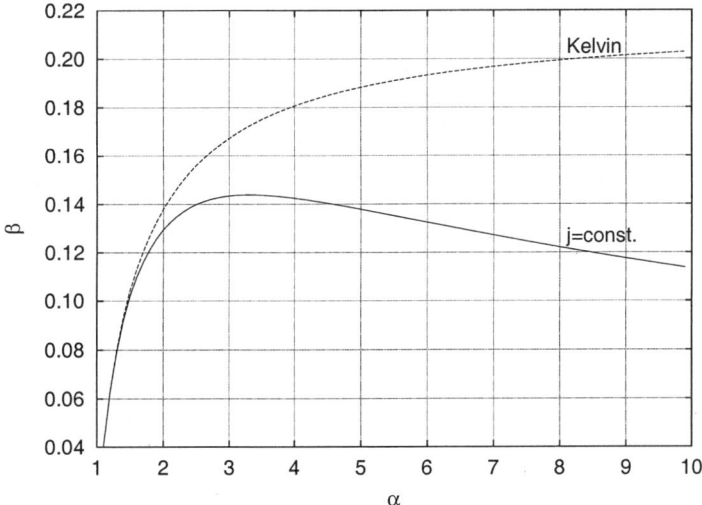

Fig. 2.11. Fabry factor of a coil with constant current density and optimal shape as a function of α, the ratio of the outermost radius to the inner radius of the coil. For comparison the Fabry factor for the same coil with the Kelvin distribution is also shown. The optimal Kelvin coil excels for $\alpha > 3.0$ the best Bitter coil ($G_{\max} = 0.1665$) and for $\alpha > 4.6$ the best Gaume coil ($G_{\max} = 0.1851$). The maximum possible Fabry factor is achieved with an optimal Kelvin coil in the limit of $\alpha \to \infty$ and has the value $\frac{1}{\sqrt{6\pi}} \approx 0.2303$

$$L = \frac{\mu_0}{4\pi} a_1 \Lambda(\alpha, \beta) \,. \tag{2.34}$$

Hence, we find for the time constant:

$$\tau = \frac{\mu_0}{8\pi^2} \frac{\lambda a_1^2}{\varrho_0} \Lambda(\alpha, \beta) \frac{\left(\int_1^\alpha du \int_{-\beta}^\beta dv\, f\right)^2}{\int_1^\alpha du \int_{-\beta}^\beta dv\, u f^2} \,. \tag{2.35}$$

The time constant is determined if the geometrical dimensions of the coil (a_1, cross-section, $\Lambda(\alpha,\beta)$), the type of current distribution (f, λ) and the conductor material (ϱ_0) are given. In particular, it is independent of the number of turns, N, which allows us to adjust for a given time constant the coil's inductance to any desired value.

As we progress along the series of coil with constant current density, Bitter coil, Gaume coil and Kelvin coil, more and more elaborate distributions of the current density are incorporated. For the corresponding Fabry factor ever-higher values are possible, which means that for the same power ever-higher fields can be generated.

Instead of varying the current density in the rectangular cross-section of the coil one can also vary the cross-section. This is demonstrated for a coil with constant current density. However, the optimal shape ('apple') in the case of constant current densities results only in minor improvements of the Fabry factor.

From a manufacturing point of view the wire-wound coil as well as the Bitter and Gaume coils are fairly easy to fabricate; for the Kelvin distribution and coils with the optimal shape the so-called polyhelix technique seems appropriate.

The Fabry formula yields an expression for the field at the center of the solenoid as a function of the geometric dimensions (Fabry factor $G(\alpha, \beta)$), the conductor material (resistivity ϱ_0), the filling factor λ, the inner bore a_1 and the power P fed into the magnet.

The magnetic field at points off the center of the coil is equally important to know, however. From an experimental point of view the field not only at the center but in the whole probe volume is of interest; usually a certain homogeneity is required or a certain field gradient is desired.

From a constructor's point of view the field in the coil is necessary to be known for calculating the Lorentz forces. The mechanical stresses induced by the Lorentz forces as well as cooling considerations may lead to different coil shapes and current distributions in an optimally designed magnet. Therefore, methods are needed to calculate the field at any point in space. This issue will be addressed in the next chapter.

In the following subsections we summarize the formulas for the Fabry factor, the resistance and the time constant for the coil with constant current density, the Bitter coil, the Gaume coil and the Kelvin coil. A comparison of Fabry factors for these coil types is shown in Table 2.1.

Homogeneous Coil

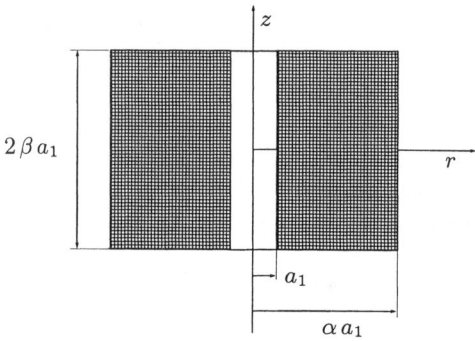

Fig. 2.12. Coil with constant current density

$$f = 1$$

$$G = \sqrt{\frac{\beta}{2\pi(\alpha^2 - 1)}} \ln \frac{\alpha + \sqrt{\alpha^2 + \beta^2}}{1 + \sqrt{1 + \beta^2}}$$

$$R = N^2 \frac{\varrho_0}{\lambda a_1} \frac{\pi(\alpha + 1)}{2(\alpha - 1)\beta}$$

$$\tau = \frac{\mu_0}{4\pi} \frac{\lambda a_1^2}{\varrho_0} \frac{2\beta(\alpha - 1)}{\pi(\alpha + 1)} \Lambda(\alpha, \beta)$$

Bitter Coil

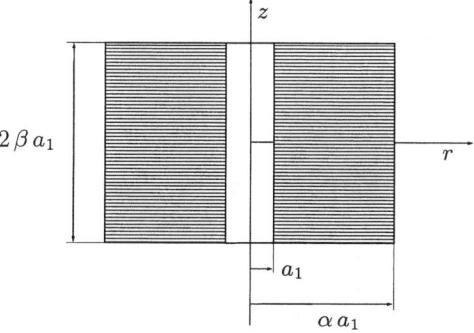

Fig. 2.13. Bitter coil

$$f = \frac{a_1}{r}$$

$$G = \frac{1}{\sqrt{4\pi\beta \ln \alpha}} \ln \frac{\alpha\left(\beta + \sqrt{1 + \beta^2}\right)}{\beta + \sqrt{\alpha^2 + \beta^2}}$$

$$R = N^2 \frac{\varrho_0}{\lambda a_1} \frac{\pi}{\beta \ln \alpha}$$

$$\tau = \frac{\mu_0}{4\pi} \frac{\lambda a_1^2}{\varrho_0} \frac{\beta \ln \alpha}{\pi} \Lambda(\alpha, \beta)$$

Gaume Coil

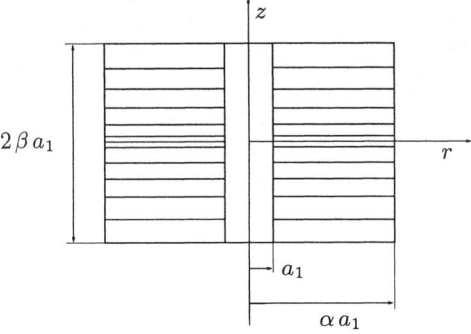

Fig. 2.14. Gaume coil

$$f = \frac{\alpha}{\alpha-1}\frac{a_1}{r}\left[\frac{1}{\sqrt{1+\left(\frac{z}{a_1}\right)^2}} - \frac{1}{\sqrt{\alpha^2+\left(\frac{z}{a_1}\right)^2}}\right]$$

$$G = \frac{1}{\sqrt{4\pi\alpha\ln\alpha}}\sqrt{\alpha\arctan\beta + \arctan\frac{\beta}{\alpha} - 2F\left(\arctan\beta, \sqrt{\frac{\alpha^2-1}{\alpha^2}}\right)}$$

$$R = N^2\frac{\varrho_0}{\lambda a_1}\frac{\pi\left[\alpha\arctan\beta + \arctan\frac{\beta}{\alpha} - 2F\left(\arctan\beta, \sqrt{\frac{\alpha^2-1}{\alpha^2}}\right)\right]}{\alpha\ln\alpha\left[\ln\frac{\alpha\left(\beta+\sqrt{1+\beta^2}\right)}{\beta+\sqrt{\alpha^2+\beta^2}}\right]^2}$$

$$\tau = \frac{\mu_0}{4\pi}\frac{\lambda a_1^2}{\varrho_0}\frac{\alpha\ln\alpha\left[\ln\frac{\alpha\left(\beta+\sqrt{1+\beta^2}\right)}{\beta+\sqrt{\alpha^2+\beta^2}}\right]^2}{\pi\left[\alpha\arctan\beta + \arctan\frac{\beta}{\alpha} - 2F\left(\arctan\beta, \sqrt{\frac{\alpha^2-1}{\alpha^2}}\right)\right]}\Lambda(\alpha,\beta)$$

Kelvin Distribution

$$f = \frac{a_1^2 r}{(r^2+z^2)^{3/2}}$$

$$G = \frac{1}{\sqrt{32\pi}}\sqrt{3\left(\arctan\beta - \frac{1}{\alpha}\arctan\frac{\beta}{\alpha}\right) + \beta\left(\frac{1}{1+\beta^2} - \frac{1}{\alpha^2+\beta^2}\right)}$$

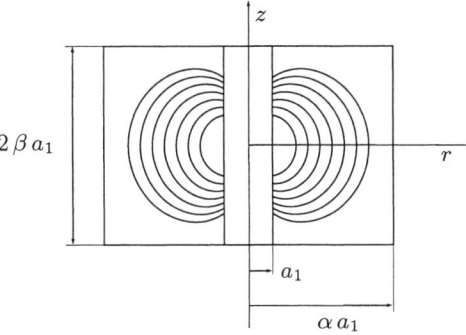

Fig. 2.15. Kelvin coil

$$R = N^2 \frac{\varrho_0}{\lambda a_1} \frac{\pi}{8} \frac{3\left[\arctan\beta - \frac{1}{\alpha}\arctan\frac{\beta}{\alpha}\right] + \beta\left[\frac{1}{1+\beta^2} - \frac{1}{\alpha^2+\beta^2}\right]}{\left[\ln\frac{\alpha\left(\beta+\sqrt{1+\beta^2}\right)}{\beta+\sqrt{\alpha^2+\beta^2}}\right]^2}$$

$$\tau = \frac{\mu_0}{4\pi} \frac{\lambda a_1^2}{\varrho_0} \frac{8}{\pi} \frac{\left[\ln\frac{\alpha\left(\beta+\sqrt{1+\beta^2}\right)}{\beta+\sqrt{\alpha^2+\beta^2}}\right]^2}{3\left[\arctan\beta - \frac{1}{\alpha}\arctan\frac{\beta}{\alpha}\right] + \beta\left[\frac{1}{1+\beta^2} - \frac{1}{\alpha^2+\beta^2}\right]} \Lambda(\alpha,\beta)$$

Comparison of Fabry Factors

Table 2.1. Comparison of the Fabry factors of coils with a rectangular cross-section for different distributions of the current density

Coil type	Fabry Factor	Condition	
Wire-wound	0.1426	$\alpha = 3.095$	$\beta = 1.862$
Bitter coil	0.1665	$\alpha = 6.423$	$\beta = 2.146$
Gaume coil	0.1854	$\alpha = 7.762$	$\beta \to \infty$
	0.1851	$\alpha = 8.000$	$\beta = 8.000$
Kelvin coil	0.2165	$\alpha \to \infty$	$\beta \to \infty$
	0.2021	$\alpha = 8.000$	$\beta = 8.000$

54   2. Analytical Calculations

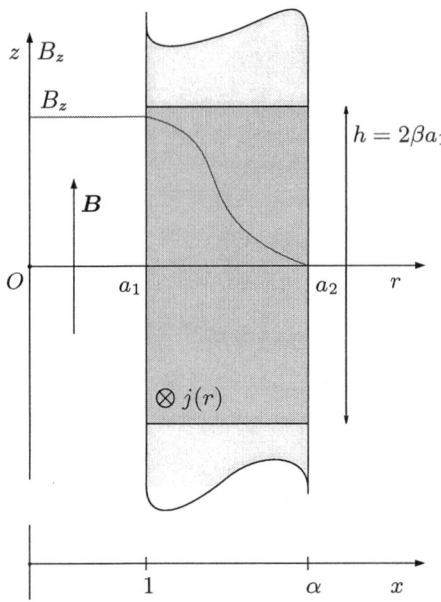

Fig. 2.16. Cylindrical coil with inner radius a_1, outer radius a_2 and height h. With the trick of extending the coil to infinity in the axial direction the field in the coil can be calculated analytically. In that case the field in the exterior region $r > a_2$ is zero. The current density $j(r)$ generates a constant field in the interior $0 \leq r < a_1$, and a varying field in the region of the coil $a_1 \leq r \leq a_2$. The current density has only an azimuthal component, and the field has only axial components. It is very convenient to express all radii in units of the inner radius a_1. We define then the outer radius with α and use the symbol x for an arbitrary radius

2.2 No Radial Transmission of Forces

We first develop the formalism for an infinitely long resistive coil with an arbitrary distribution of the current density. Additionally, there exists no mechanical transmission of forces in the radial direction. Then the fields, the current density and the mechanical stresses can be expressed in an analytical form. Later this formalism will be applied to some representative examples. This coil model is often referred to as a coil with free-standing turns.

The situation is sketched in Fig. 2.16. There is a cylindrical coil with inner radius a_1, outer radius a_2 and height h with an azimuthal current density $j(r)$, which may depend on the radial position. An analytical calculation of the field becomes possible if we think of the coil as extended to infinity in the axial direction. In that case the magnetic field has only an axial component. The field in the exterior region $r > a_2$ is zero, in the interior region $0 \leq r < a_1$ it is constant, and in the region of the coil, $a_1 \leq r \leq a_2$, it varies. The current density is written in the general form

$$j(r) = \lambda j_0 f(r), \tag{2.36}$$

where λ denotes the filling factor, which describes how much of the total cross-section is filled with conductor. A fraction $(1 - \lambda)$ of the cross-section is filled, for instance with insulation. The current density in the conductor is given by $j_0 f(r)$, whereas $\lambda j_0 f(r)$ gives the local average of the current density over the coil's cross-section. The dimensionless distribution function $f(r)$, which is by definition zero outside the coil, is normalized to have a maximum value of 1 at some point(s) within the coil:

$$\text{Max}(f(r), r \in [a_1, a_2]) = 1. \tag{2.37}$$

With the help of the Maxwell equation $\int H \, d\mathbf{s} = \iint \mathbf{j} \, d\mathbf{A}$ the axial field can be calculated to be

$$\begin{aligned} B(r) &= \mu_0 \int_{a_1}^{a_2} j(r) \, dr =: B_0 & \text{for} & \quad 0 \le r < a_1, \\ B(r) &= \mu_0 \int_{r}^{a_2} j(r) \, dr & \text{for} & \quad a_1 \le r < a_2, \\ B(r) &= 0 & \text{for} & \quad a_2 \le r. \end{aligned} \tag{2.38}$$

By introducing the normalized values for the outer radius a_2 and the radial coordinate r as

$$\alpha := \frac{a_2}{a_1} \quad \text{and} \quad x := \frac{r}{a_1} \tag{2.39}$$

we can define a function $F(x)$ as

$$F(x) := \int_{x}^{\alpha} f(u) \, du. \tag{2.40}$$

Like $f(x)$ the function $F(x)$ is a dimensionless number. Because $f(u) \equiv 0$ outside the coil, i.e. for $u \notin [1; \alpha]$, we see that

$$\begin{aligned} F(x) &= \int_{1}^{\alpha} f(u) \, du = F(1) & \text{for} & \quad 0 \le x < 1, \\ F(x) &= \int_{x}^{\alpha} f(u) \, du & \text{for} & \quad 1 \le x < \alpha, \\ F(x) &= 0 & \text{for} & \quad \alpha \le x. \end{aligned} \tag{2.41}$$

Therefore, we can write for the axial field

$$B(x) = \mu_0 \lambda j_0 a_1 F(x). \tag{2.42}$$

The central field is

$$B_0 = \mu_0 \lambda j_0 a_1 F(1). \tag{2.43}$$

2. Analytical Calculations

Because $F(x)$ is a number, the term $\mu_0 \lambda j_0 a_1$ has the dimension of a field; we will use that later in (2.55). We will later need expressions for the magnetic energy and the ohmic power in a coil. Using the model of the infinite long coil we make the following assumption for the magnetic energy: it shall be the magnetic energy of the infinite long coil, which lies in a part of axial length h. We find then for the magnetic energy of the coil

$$W_m = \frac{1}{2\mu_0} \int B^2 \, dV = \frac{\pi}{\mu_0} h \int_0^{a_2} r \, B^2(r) \, dr \tag{2.44}$$

and with (2.42) we get

$$W_m = \frac{B_0^2}{2\mu_0} \pi h \, a_1^2 \left[1 + \frac{2}{F^2(1)} \int_1^\alpha x \, F^2(x) \, dx \right] . \tag{2.45}$$

The first expression in (2.45) is the magnetic energy density in the center of the coil, the second one is the volume of the inner bore of the coil with inner radius a_1 and height h. The expression in the braces is a numerical factor reflecting the type of current density distribution and the normalized outer radius α of the coil.

For calculating the the ohmic heat generated in the coil we have to integrate ϱj^2 over the coil volume, where ϱ is the local average resistivity and j the local average current density in the coil. Using the resistivity ϱ_0 of the conductor, we must substitute

$$\varrho = \frac{1}{\lambda} \varrho_0 \tag{2.46}$$

and obtain for the generated ohmic power P_{ohm} the expression

$$P_{\text{ohm}} = \varrho_0 \, j_0^2 \, \lambda \, 2\pi \, h \int_{a_1}^{a_2} r \, f^2(r) \, dr \tag{2.47}$$

and with (2.42) we obtain the expression

$$P_{\text{ohm}} = \frac{B_0^2}{2\mu_0} \pi h a_1^2 \, \frac{\varrho_0}{\lambda \mu_0} \frac{1}{a_1^2} \frac{4}{F^2(1)} \int_1^\alpha x \, f^2(a_1 x) \, dx . \tag{2.48}$$

As above, the first two expressions in (2.48) are the magnetic energy density at the center and the volume of the inner bore. The remainder describes the type of the coil (f, F, a_1 and α) and the used conductor (resistivity ϱ_0 and filling factor λ); it has the dimension of s^{-1}.

For the general calculation of the stress tensor $\boldsymbol{\sigma}$ we have to solve the equilibrium equation

$$\nabla \boldsymbol{\sigma} + \boldsymbol{f} = 0 , \tag{2.49}$$

where $\boldsymbol{f}$ is the density of the Lorentz forces, in conjunction with the so-called stress-strain relations. Fortunately, in the case of an infinite long coil without radial transmission of mechanical forces the stress tensor has only one non-vanishing component, the azimuthal one, and the density of the Lorentz forces only a radial one, $f_r = j(r) B(r)$, so that the equilibrium condition (2.49) gets reduced to an algebraic equation

$$-\frac{1}{r} \sigma_{\varphi\varphi}(r) + j(r) B(r) = 0 , \qquad (2.50)$$

which can be rewritten with our definitions as

$$\sigma_{\varphi\varphi}(x) = \frac{1}{\mu_0} \left(\mu_0 \lambda j_0 a_1\right)^2 \, x f(x) F(x) , \quad \text{with } x \in [1; \alpha] . \qquad (2.51)$$

2.2.1 Boundary Conditions

There are two boundary conditions which limit the maximum achievable field in a coil. The first is caused by the finite mechanical strength of the coil. We will see that the type of the coil, i.e. coil with constant current density, Bitter coil, etc., will affect this boundary condition in a considerable way.

The second boundary condition is a thermal one. The current flowing through an resistive coil will generate ohmic heat. In static coils this heat has to be removed by a coolant; in pulsed coils the heat is absorbed adiabatically by the coil itself.

Finite Mechanical Strength. The boundary condition of finite mechanical strength defines a characteristic current density j_σ and a characteristic field B_σ. The starting point is (2.51):

$$\sigma_{\varphi\varphi}(x) = \frac{1}{\mu_0} \left(\mu_0 \lambda j_0 a_1\right)^2 \, x f(x) F(x) , \quad \text{with } x \in [1; \alpha] . \qquad (2.52)$$

Somewhere in the coil the azimuthal stress has a maximum value. We define the maximum of the function $x f(x) F(x)$ in the interval $x \in [1; \alpha]$ by $\mathcal{M}$:

$$\mathcal{M} := \text{Max}\left(x f(x) F(x), x \in [1; \alpha]\right) . \qquad (2.53)$$

Therefore, the maximum stress is

$$\sigma_{\max} = \frac{1}{\mu_0} \left(\mu_0 \lambda j_0 a_1\right)^2 \mathcal{M} . \qquad (2.54)$$

Noting that the expression within the brackets has the dimension of Tesla, we define a stress-determined field as (we use the new subscript σ for the current density)

$$B_\sigma = \mu_0 \lambda j_\sigma a_1 = \sqrt{\mu_0 \sigma_{\max}} \, \frac{1}{\sqrt{\mathcal{M}}} . \qquad (2.55)$$

This is the dimensional part of (2.43). Resolving for the current density yields

$$j_\sigma = \sqrt{\frac{\sigma_{\max}}{\mu_0}} \, \frac{1}{\lambda a_1} \, \frac{1}{\sqrt{\mathcal{M}}} . \qquad (2.56)$$

Finite Cooling Capability. In a static, resistive coil the ohmic heat generated by the current has to be removed by a coolant. The power density p at a point in the coil depends on the resistivity and the current density at that point, $p(r) = \varrho j^2(r)$. Again, we work with the resistivity of the conductor and find

$$p(r) = \left(\frac{\varrho_0}{\lambda}\right)(\lambda j_0 f(r))^2 = \lambda \varrho_0 j_0^2 f^2(r) \,. \tag{2.57}$$

The resistivity is assumed to be constant throughout the coil. The highest power density occurs where the function $f(r)$ has its maximum, which is defined to be 1:

$$p_{\max} = \lambda \varrho_0 j_0^2 f_{\max}^2 \quad \text{with } f_{\max} := \text{Max}\left(f(x), x \in [1; \alpha]\right) = 1\,. \tag{2.58}$$

This power has to be removed by a proper cooling design. Now a thermally defined current density can be defined by

$$j_{\text{th,static}} = \sqrt{\frac{p}{\lambda \varrho_0}} \tag{2.59}$$

as well as a thermally defined field by

$$B_{\text{th,static}} = \mu_0 \lambda j_{\text{th,static}} a_1 = \mu_0 \sqrt{\frac{\lambda p}{\varrho_0}} a_1 \,. \tag{2.60}$$

Finite Heat Capacity. In a pulsed coil the current is allowed to flow only for a short time, during which the coil will be heated adiabatically. In Sect. 1.3.1 a relation between the current density in the conductor, the pulse length, the conductor material and the initial and final temperature was developed. With the definition of the current density in this chapter of $j(r) = \lambda j_0 f(r)$ (see (2.36)) we find the maximum current density in the conductor throughout the coil's cross-section to be j_0. By introducing a new subscript $(0 \rightarrow \text{th, pulse})$ we get

$$(j_{\text{th,pulse}})^2 \, t_{\text{pulse}} \, \xi = \mathcal{F}_{\text{Mat}}(T_i, T_f) \,. \tag{2.61}$$

Here t_{pulse} is the length of the current pulse and ξ reflects the shape of the current pulse, with $\xi = 1$ for a rectangular current pulse, $\xi = \frac{1}{2}$ for a half-period of a sine wave and $\xi = \frac{1}{3}$ for a triangular waveform. The material integral $\mathcal{F}_{\text{Mat}}(T_i, T_f)$ depends on the conductor material and the initial and final temperature.

By analogy with the previous section the boundary condition of a finite heat capacity of the coil defines a characteristic current density and a characteristic field:

$$j_{\text{th,pulse}} = \sqrt{\frac{\mathcal{F}_{\text{Mat}}(T_i, T_f)}{t_{\text{pulse}} \, \xi}}\,, \tag{2.62}$$

$$B_{\text{th,pulse}} = \mu_0 \lambda a_1 j_{\text{th,pulse}} \mu_0 \lambda \sqrt{\frac{\mathcal{F}_{\text{Mat}}(T_i, T_f)}{t_{\text{pulse}} \, \xi}} a_1 \,. \tag{2.63}$$

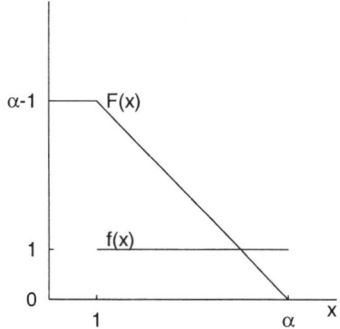

Fig. 2.17. $f(x)$ and $F(x)$ for the case of constant current density

Both Boundary Conditions. The maximum achievable field for a coil is determined by the stronger of the two aforementioned boundary conditions. For pulsed coils the conditions are finite mechanical strength and finite heat capacity. Depending on the design parameters such as the strength of the coil, coil geometry, distribution of the current density, temperature range and desired pulse length, either the finite mechanical strength or the finite heat capacity of the coil will be the stronger condition. For static coils one has to consider both the finite mechanical strength and the finite cooling capability.

In any case, the field of the coil can be written in the general form

$$B(x) = \text{Min}(\,B_{\text{th}}, B_\sigma\,)\, F(x)\,, \tag{2.64}$$

where B_{th} stands either for the characteristic field of finite heat capacity or finite cooling capability. In the case $B_{\text{th}} < B_\sigma$ the coil is governed by the thermal boundary condition, in the other case of $B_\sigma < B_{th}$ it is the finite mechanical strength which limits the maximum achievable field.

The formalism developed here will now be applied to some representative current distributions. The three coil types – solenoid with constant current density, Bitter coil and stress-optimized coil – will be treated separately.

2.2.2 Coil with Constant Current Density

The current density is constant over the whole cross-section:

$$j(r) = \lambda j_0 = \lambda j_\sigma = \text{const.} \tag{2.65}$$

Therefore (see (2.36) and (2.40)), we get for the two dimensionless functions $f(x)$ and $F(x)$ (see Fig. 2.51):

$$f(x) = 1 \quad \text{and} \quad F(x) = \begin{cases} \alpha - 1 & \text{for } 0 \leq x < 1 \\ \alpha - x & \text{for } 1 \leq x < \alpha \\ 0 & \text{for } \alpha \leq x \end{cases}. \tag{2.66}$$

The characteristic current density and field from the boundary condition of finite stress capability are then

$$j_\sigma = \sqrt{\frac{\sigma_{\max}}{\mu_0}} \frac{1}{\lambda a_1} \begin{cases} \frac{2}{\alpha} & \text{for } 2 \leq \alpha \\ \sqrt{\frac{1}{\alpha-1}} & \text{for } 1 < \alpha < 2 \end{cases}, \tag{2.67}$$

$$B_\sigma = \sqrt{\mu_0 \sigma_{\max}} \begin{cases} \frac{2}{\alpha} & \text{for } 2 \leq \alpha \\ \sqrt{\frac{1}{\alpha-1}} & \text{for } 1 < \alpha < 2 \end{cases}. \tag{2.68}$$

For the central field we find

$$B_0 = \operatorname{Min}(B_{\text{th}}, B_\sigma)(\alpha - 1) \tag{2.69}$$

with B_{th} as the characteristic thermal field, which is derived from the finite heat capacity in the pulsed case and from the finite cooling capability in the static case.

The magnetic energy associated with the coil becomes (remember the assumption that it is the magnetic energy of the infinitely long coil associated with a part of axial length h)

$$W_{\text{m}} = \frac{B_0^2}{2\mu_0} \pi h a_1^2 \frac{\alpha^2 + 2\alpha + 3}{6} \tag{2.70}$$

and for the ohmic power we find

$$P_{\text{ohm}} = \frac{B_0^2}{2\mu_0} \pi h a_1^2 \frac{\varrho_0}{\lambda \mu_0} \frac{1}{a_1^2} 2 \frac{\alpha+1}{\alpha-1}. \tag{2.71}$$

2.2.3 Bitter Coil

In a Bitter coil the current density falls off with the inverse radius as

$$j(x) = \lambda j_0 \frac{1}{x}, \tag{2.72}$$

from which we find the two dimensionless functions $f(x)$ and $F(x)$ (see Fig. 2.18) to be

$$f(x) = \frac{1}{x} \quad \text{and} \quad F(x) = \begin{cases} \ln \alpha & \text{for } 0 \leq x < 1 \\ \ln \frac{\alpha}{x} & \text{for } 1 \leq x < \alpha \\ 0 & \text{for } \alpha \leq x \end{cases}. \tag{2.73}$$

The boundary condition of finite stress results in the following characteristic current density and characteristic field:

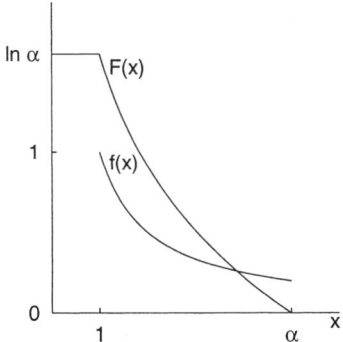

Fig. 2.18. $f(x)$ and $F(x)$ for a Bitter coil

$$j_\sigma = \sqrt{\frac{\sigma_{\max}}{\mu_0}} \frac{1}{\lambda a_1} \frac{1}{\sqrt{\ln \alpha}}, \tag{2.74}$$

$$B_\sigma = \sqrt{\mu_0 \sigma_{\max}} \frac{1}{\sqrt{\ln \alpha}}. \tag{2.75}$$

We find for the central field

$$B_0 = \text{Min}(B_{\text{th}}, B_\sigma) \ln \alpha, \tag{2.76}$$

with the characteristic thermal field defined as above. The magnetic energy associated with a coil piece of axial length h is calculated using (2.45) to be

$$W_{\text{m}} = \frac{B_0^2}{2\mu_0} \pi h a_1^2 \frac{\alpha^2 - 1 - 2\ln\alpha}{2(\ln \alpha)^2}, \tag{2.77}$$

and for the ohmic power we find (see (2.48))

$$P_{\text{ohm}} = \frac{B_0^2}{2\mu_0} \pi h a_1^2 \frac{\varrho_0}{\lambda \mu_0} \frac{1}{a_1^2} \frac{4}{\ln \alpha}. \tag{2.78}$$

2.2.4 Stress-Optimized Coil

The optimization aims at having constant mechanical stresses throughout the coil, which would utilize the mechanical strength of the conductor in the most effective way. In contrast to the former coils we allow for the moment a non-zero outer field B_{a}.

The mechanical stress has only an azimuthal component, which can be written in the form (see (2.42))

$$\sigma_{\varphi\varphi} = r\, j(r)\, B(r) = -\frac{1}{\mu_0} r\, B(r) \frac{dB}{dr}, \tag{2.79}$$

and we have to find a solution for $B(r)$ which leaves $\sigma_{\varphi\varphi}$ constant. We find for the field and the current density in the coil region

2. Analytical Calculations

$$B(r) = \sqrt{B_a^2 + 2\mu_0 \sigma_{\varphi\varphi} \ln \frac{a_2}{r}}, \tag{2.80}$$

$$j(r) = \sigma_{\varphi\varphi} \frac{1}{r\sqrt{B_a^2 + 2\mu_0 \sigma_{\varphi\varphi} \ln \frac{a_2}{r}}}. \tag{2.81}$$

Hence the current density at the outer radius a_2 is

$$j(a_2) = \frac{\sigma_{\varphi\varphi}}{a_2 B_a}. \tag{2.82}$$

This equation states that an outer field $B_a \neq 0$ is necessary in order to avoid $j(a_2)$ from diverging. One can show that otherwise the pulse length of a pulsed magnet would become zero (see (2.61)).

The general form of the current density as a function of the radii is a valley with maxima at the inner and outer radii. The bigger one of these local maxima determines the pulse length for a pulsed magnet (see (2.61) and Fig. 2.19).

With a proper chosen outer field B_a one can make the current density have the same value at the inner and the outer radii of the coil. This defines B_a as

$$j(a_1) = j(a_2) \implies B_a = \sqrt{\mu_0 \sigma_{\varphi\varphi}} \sqrt{\frac{2\ln\alpha}{\alpha^2 - 1}} \quad \text{with } \alpha := \frac{a_2}{a_1}. \tag{2.83}$$

This outer field is generated by a coil with constant current density extending in the radial direction from a_2 to a_3. This means the total coil is subdivided into an inner part, which is optimized due to constant stress, and an outer coil with constant current density. The field disappears for $r > a_3$. In Fig. 2.20 the current density, the field and the stress of an optimized coil are drawn schematically.

From the current density we can now identify the dimensionless function $f(x)$ (see Fig. 2.21) as

$$f(x) = \frac{\alpha}{x} \frac{1}{\sqrt{1 + (1 - \frac{\ln x}{\ln \alpha})(\alpha^2 - 1)}}, \tag{2.84}$$

defined for $x \in [1, \alpha]$. Throughout the outer coil we have $f(x) = 1$. We calculate $F(x)$ (see Fig. 2.21) in the inner coil to be

$$F(x) = \frac{2\alpha \ln \alpha}{\alpha^2 - 1} \sqrt{1 + \frac{\alpha^2 - 1}{\ln \alpha} \ln \frac{\alpha}{x}} \quad \text{for } x \in [1, \alpha]. \tag{2.85}$$

In the outer coil the function $F(x)$ decreases linearly to 0. The boundary condition of finite mechanical strength leads to the following characteristic current densities and fields:

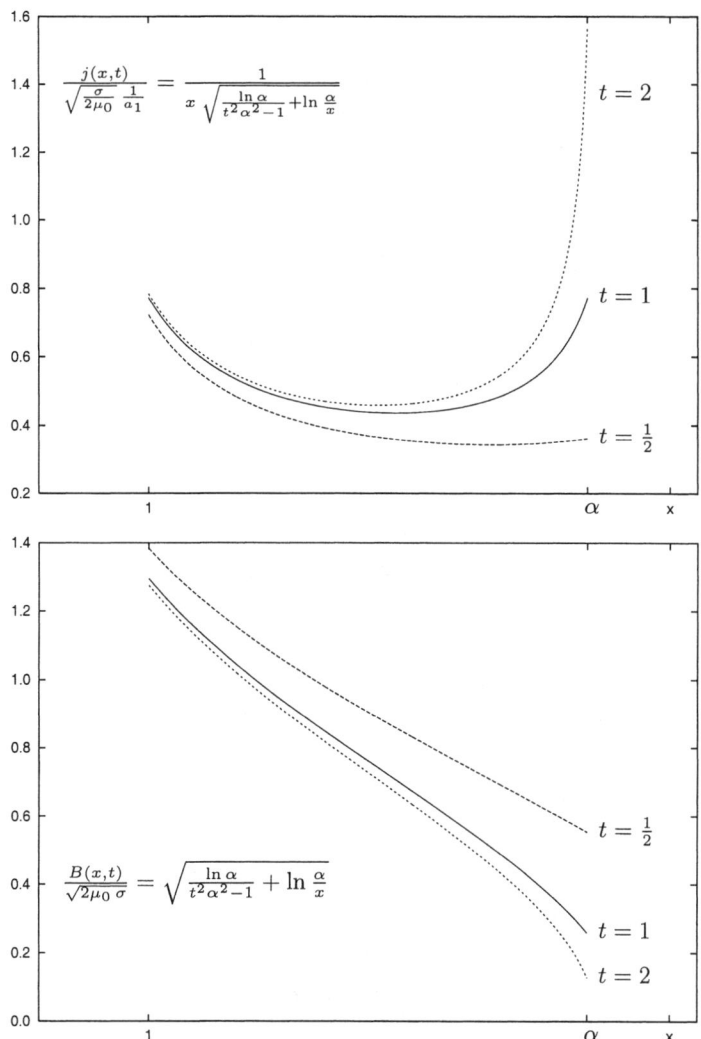

Fig. 2.19. Normalized current density and field for an infinite, stress-optimized coil. An additional parameter t has been introduced here, defined as the ratio of the current density at the outer coil side ($x = \alpha$ in relative units) and the inner side ($x = 1$): $j(\alpha) = t \cdot j(1)$. The current density for every parameter t leads to a constant stress distribution within the coil, there are differences in the pulse length and the supporting outer field, however. The pulse length is determined by the maximum current density in the coil (see (2.61)); depending on the parameter t this happens to be the current density on either the inner ($t < 1$) or the outer side ($t > 1$) of the coil. Hence, for long pulse lengths one should use parameters $t < 1$. On the other hand, this leads to increasingly high values of the necessary outer field, as can be seen in the lower graph. In the text we choose therefore $t = 1$

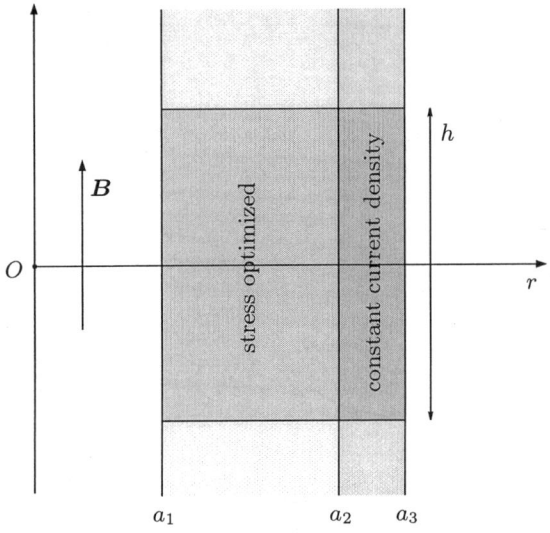

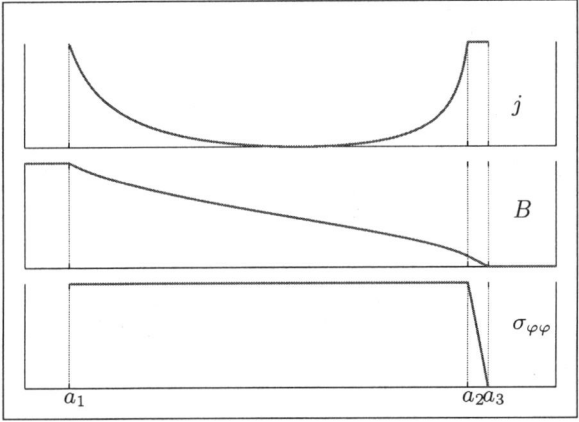

Fig. 2.20. Situation for a stress-optimized coil, consisting of an inner part extending from a_1 to a_2 and an outer part from a_2 to a_3. Again the trick of extending the coil to infinity in the axial direction is used in order to calculate the fields and stresses. The inner part is optimized with respect to constant stress, and the outer part is a homogeneous coil (constant current density). In the lower box the current density j, the field B and the stress $\sigma_{\varphi\varphi}$ are shown as a function of the radial position

$$j_\sigma = \sqrt{\frac{\sigma_{\max}}{\mu_0}} \frac{1}{\lambda a_1} \sqrt{\frac{\alpha^2 - 1}{2\alpha^2 \ln \alpha}}, \tag{2.86}$$

$$B_\sigma = \sqrt{\mu_0 \sigma_{\max}} \sqrt{\frac{\alpha^2 - 1}{2\alpha^2 \ln \alpha}}. \tag{2.87}$$

For the central field we find

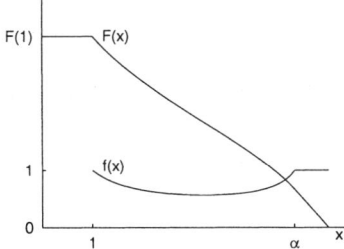

Fig. 2.21. The two functions $f(x)$ and $F(x)$ for the stress optimized coil (see (2.84) and (2.85)). The coil occupies the region $[1, \alpha]$ in relative units of the inner radius a_1

$$B_0 = B_\sigma F(1) = \sqrt{\mu_0 \sigma_{\max}} \sqrt{\frac{2\alpha^2 \ln \alpha}{\alpha^2 - 1}} \qquad (2.88)$$

and the magnetic energy in the inner coil up to the radius a_2 is calculated to be

$$W_{\mathrm{m}} = \frac{B_0^2}{2\mu_0} \pi h a_1^2 \left[\frac{(\alpha^2 - 1)^2}{2\alpha^2 \ln \alpha} + 1 \right]. \qquad (2.89)$$

The stress in the inner part of the coil is constant and $\sigma_{\varphi\varphi} = \sigma_{\max}$.

The Outer Coil The outer coil has to produce the field B_a in its interior, and the field on the outside has to disappear. We use the results from Sect. 2.2.2 and denote the corresponding quantities for the outer coil by a tilde ($\alpha \to \tilde{\alpha}$, $B_\sigma \to \tilde{B}_\sigma$).

Case A: Finite mechanical stress and $\tilde{\alpha} > 2$.
The inner field is

$$B_\mathrm{a} = \tilde{B}_\sigma (\tilde{\alpha} - 1) \quad \text{with} \quad \tilde{B}_\sigma = \sqrt{\mu_0 \sigma_{\max}} \, \frac{2}{\tilde{\alpha}} . \qquad (2.90)$$

Resolving for $\tilde{\alpha}$ leads to

$$\tilde{\alpha} = \frac{1}{1 - \frac{1}{2}\sqrt{\frac{2\ln\alpha}{\alpha^2-1}}} < 2 \quad \text{for} \quad \alpha > 1. \qquad (2.91)$$

Therefore, this definition causes a contradiction to the preposition $\tilde{\alpha} > 2$.

Case B: Finite mechanical stress and $1 < \tilde{\alpha} <= 2$.
Now we have

$$B_\mathrm{a} = \tilde{B}_\sigma (\tilde{\alpha} - 1) \quad \text{and} \quad \tilde{B}_\sigma = \sqrt{\mu_0 \sigma_{\max}} \, \frac{1}{\sqrt{\tilde{\alpha} - 1}}, \qquad (2.92)$$

from which follows

$$\tilde{\alpha} = 1 + \frac{2 \ln \alpha}{\alpha^2 - 1} < 2 \quad \text{for} \quad \alpha > 1 \,. \tag{2.93}$$

This is a self-consistent solution. The additional contribution W_{add} to the magnetic energy is calculated to be

$$W_{\text{add}} = \frac{B_0^2}{2\mu_0} \pi h a_2^2 \, \frac{\tilde{\alpha}^2 + 2\tilde{\alpha} - 3}{6} \tag{2.94}$$

$$= \frac{B_0^2}{2\mu_0} \pi h a_2^2 \left[\frac{1}{6} \frac{B_a^2}{\mu_0 \sigma_{\max}} \left(4 + \frac{B_a^2}{\mu_0 \sigma_{\max}} \right) \right] . \tag{2.95}$$

Inner and Outer Coil The stress-optimized coil consists of an inner coil with constant mechanical stresses and an outer coil producing the necessary supporting field B_a for the inner coil. For simplicity the outer coil was chosen as a coil with constant current density.

The magnetic energy of the whole coil is found by adding the two expressions for the magnetic energy of the inner and the outer coils (see (2.89), (2.95) and (2.83)):

$$W_{\text{m}} = \frac{B_0^2}{2\mu_0} \pi h a_1^2 \left[\frac{(\alpha^2 - 1)^2}{2\alpha^2 \ln \alpha} + 1 + \frac{2}{3} \frac{\ln \alpha}{\alpha^2 - 1} \left(2 + \frac{\ln \alpha}{\alpha^2 - 1} \right) \right] . \tag{2.96}$$

2.2.5 Comparison of the Three Coil Types

The Central Field (Fig. 2.23). We look first at the behavior of the different fields derived from the two different boundary conditions (see Fig. 2.22). The field associated with the finite heat capacity is

$$B_{\text{th}} = \mu_0 \lambda \, a_1 \sqrt{\frac{\mathcal{F}_{\text{Mat}}(T_i, T_f)}{t_{\text{pulse}} \, \xi}} > 0 \,. \tag{2.97}$$

It is finite, positive and independent of the coil parameter α.

In the case of large α the fields caused by the finite mechanical strength for the coil with constant current density ('con'), the Bitter coil ('bit') and the stress-optimized coil ('opt') are

$$B_{\sigma,\text{con}} = \sqrt{\mu_0 \sigma_{\max}} \, \frac{2}{\alpha} \quad \to 0 \quad \text{for} \quad \alpha \to \infty \,, \tag{2.98}$$

$$B_{\sigma,\text{bit}} = \sqrt{\mu_0 \sigma_{\max}} \, \frac{1}{\sqrt{\ln \alpha}} \quad \to 0 \quad \text{for} \quad \alpha \to \infty \,, \tag{2.99}$$

$$B_{\sigma,\text{opt}} = \sqrt{\mu_0 \sigma_{\max}} \sqrt{\frac{\alpha^2 - 1}{2\alpha^2 \ln \alpha}} \quad \to 0 \quad \text{for} \quad \alpha \to \infty \,. \tag{2.100}$$

They all converge to 0 for $\alpha \to \infty$.

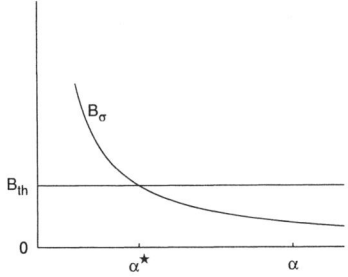

Fig. 2.22. Behavior of the fields associated to the boundary conditions of finite heat capacity, B_{th}, and of finite mechanical strength, B_σ, as a function of the coil parameter α

This means that there is a certain α^*, above which the respective coil type is governed by the finite mechanical strength, $B_\sigma < B_{\text{th}}$. For the region below α^* the coil is limited by the coil heat capacity, $B_{\text{th}} < B_\sigma$.

As a consequence one can distinguish between two optimization strategies. For small coils one has to comply with the finite heat capacity by choosing a constant current density distribution and an optimal conductor material. Large coils have to be optimal in using the finite strength of the conductor material in the best way, even at the cost of increasing the necessary power and/or energy demands.

If we ask for the maximum possible field, then we use first the boundary condition of finite mechanical stress only. The resulting pulse time t_{pulse} follows then from (2.97), combined with the condition $B_{\text{th}} = B_\sigma$. Following this path we find for the central fields in the case of large α:

$$B_{0,\text{con}} = \sqrt{\mu_0 \sigma_{\text{max}}} \; 2 \frac{\alpha - 1}{\alpha}, \tag{2.101}$$

$$B_{0,\text{bit}} = \sqrt{\mu_0 \sigma_{\text{max}}} \; \sqrt{\ln \alpha}, \tag{2.102}$$

$$B_{0,\text{opt}} = \sqrt{\mu_0 \sigma_{\text{max}}} \; \sqrt{\frac{2\alpha^2 \ln \alpha}{\alpha^2 - 1}}. \tag{2.103}$$

These functions can be written in the general form of

$$\boxed{B_0 = \sqrt{\mu_0 \sigma_{\text{max}}} \; \mathcal{B}(\alpha)}. \tag{2.104}$$

The central field depends on the yield strength σ_{max} of the conductor and a function $\mathcal{B}$, which depends only on the coil type and the coil parameter α. The three functions are shown in Fig. 2.23. For $\alpha \to \infty$ the central field of the coil with the constant current density becomes

$$B_{0,\text{hom}} = \sqrt{\mu_0 \sigma_{\text{max}}} \; 2 \quad \text{for} \quad \alpha \to \infty. \tag{2.105}$$

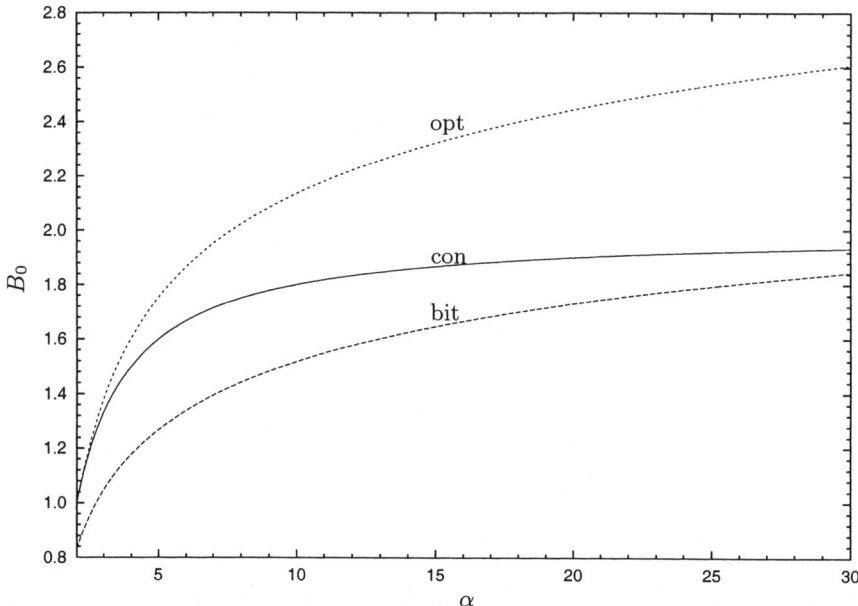

Fig. 2.23. The central field B_0 in units of $\sqrt{\mu_0 \sigma_{\max}}$ as a function of the parameter α for the three coil types: the solenoid with constant current density ('con'), the Bitter coil ('bit') and the stress-optimized coil ('opt'). The field function of the solenoid with constant current density converges to 2; the other two increase monotonically with α. The function for the optimized coil is always greater than that for the Bitter coil

This means that the central field is solely determined by the yield strength of the conductor, no matter how large one makes the magnet. The situation is different for the Bitter coil and the stress-optimized coil. One finds for the latter:

$$B_{0,\text{Bit}} = \sqrt{\mu_0 \sigma_{\max}} \sqrt{\ln \alpha} \quad \to \infty \quad \text{for} \quad \alpha \to \infty, \tag{2.106}$$

$$B_{0,\text{Opt}} = \sqrt{\mu_0 \sigma_{\max}} \sqrt{2 \ln \alpha} \quad \to \infty \quad \text{for} \quad \alpha \to \infty. \tag{2.107}$$

Here the central field depends not only on the yield strength of the conductor, but also on the coil parameter α. Additionally, the stress-optimized coil yields for the same α higher fields than the Bitter coil by a factor of $\sqrt{2}$.

We conclude: by increasing the outer radius of a coil with a stress-optimized current density distribution one can produce with a conductor of finite strength infinitely high magnetic fields.

The Magnetic Energy (Fig. 2.24). The equations for the magnetic energy – (2.70), (2.77) and (2.96) – can be written in the general form of

$$\boxed{W_m = \pi \sigma_{\max} a_1^3 \, \mathcal{W}(\alpha, \beta)}. \tag{2.108}$$

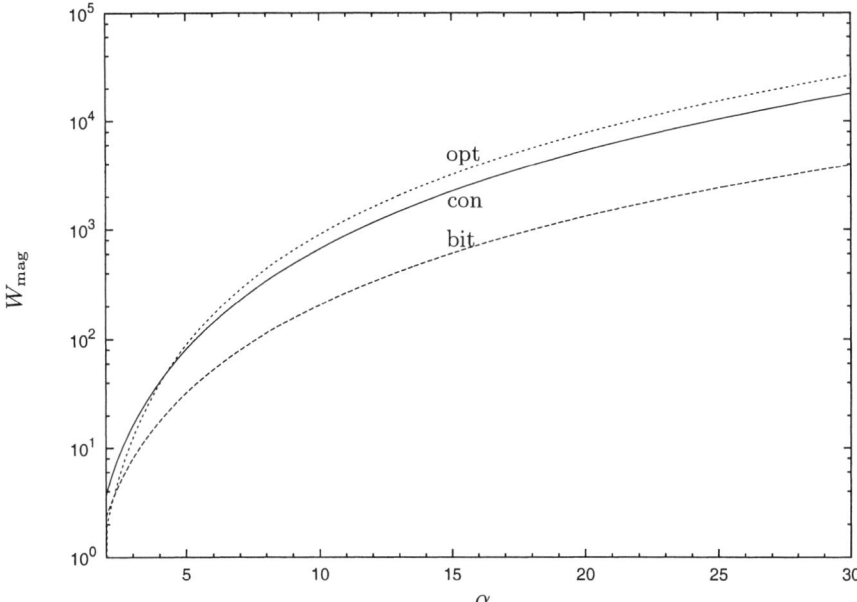

Fig. 2.24. Magnetic energy W_{mag} in units of $\pi \sigma_{max} a_1^3$ as a function of the parameter α for the three coil types: the infinite long coil with constant current density ('con'), the Bitter coil ('bit') and the stress-optimized coil ('opt'). Additionally we have used always $\alpha = \beta$

The magnetic energy is a function of the yield strength σ_{max}, the inner radius a_1 and a function $\mathcal{W}$ describing the coil type and its geometry. The leading terms in the case of large α are

$$\mathcal{W}_{con} = \frac{2}{3} \beta \alpha^2 , \tag{2.109}$$

$$\mathcal{W}_{bit} = \frac{\beta \alpha^2}{2 \ln \alpha} , \tag{2.110}$$

$$\mathcal{W}_{opt} = \beta \alpha^2 . \tag{2.111}$$

All three expressions increase with increasing α and β. The slowest increase occurs with the bitter coil; the solenoid with constant current density and the stress-optimized coil differ only by a factor of $2/3$.

The Pulse Length (Fig. 2.25). From the boundary condition of finite heat capacity of the coil we got a relation between the characteristic field B_{th} and the pulse length t_{pulse} (see (2.97)). By setting this field equal to the characteristic field of the finite stress capability, $B_\sigma = B_{th}$, we find an expression for the maximum possible pulse length:

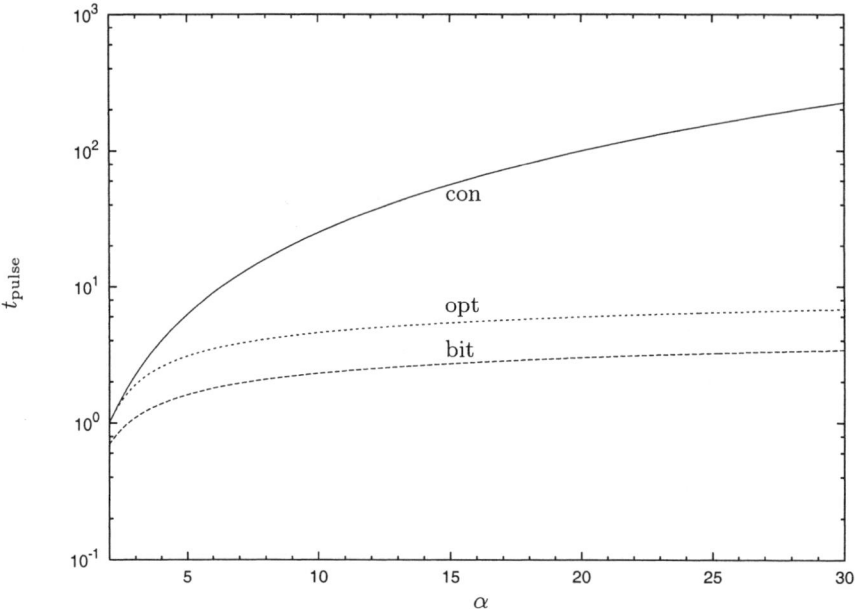

Fig. 2.25. Pulse length t_{pulse} in units of $(\mu_0 \lambda a_1)^2 \frac{\mathcal{F}_{\text{Mat}}(T_i, T_f)}{\xi \, \mu_0 \sigma_{\max}}$ as a function of the parameter α for the three infinitely long coil types. The coil with constant current density is denoted by 'con', the Bitter coil by 'bit' and the stress-optimized coil by 'opt'

$$\boxed{t_{\text{pulse}} = \frac{(\mu_0 \lambda a_1)^2}{\mu_0 \sigma_{\max}} \frac{\mathcal{F}_{\text{Mat}}(T_i, T_f)}{\xi} \mathcal{T}(\alpha)}. \tag{2.112}$$

The pulse length is given by the the filling factor λ, the inner radius a_1, the material integral $\mathcal{F}_{\text{Mat}}$, a factor ξ describing the pulse shape, the yield strength $\sigma_{\max}$ and a function $\mathcal{T}$, which depends only on the coil type and the coil parameter α. For the three coil types this function is

$$\mathcal{T}_{\text{con}} = \frac{\alpha^2}{4} \quad (\text{for} \quad 2 \leq \alpha), \tag{2.113}$$

$$\mathcal{T}_{\text{bit}} = \ln \alpha, \tag{2.114}$$

$$\mathcal{T}_{\text{opt}} = 2 \ln \alpha. \tag{2.115}$$

In the case of large α the solenoid with constant current density gives the longest pulse lengths; for the Bitter coil the pulse length increases with the logarithm of α and the stress-optimized coil is two times better.

The Power (Fig. 2.26). An estimation of the necessary power is made by dividing the magnetic energy by the half-pulse length. This is reasonable because one wants to reach the maximum of the field in a sufficiently small

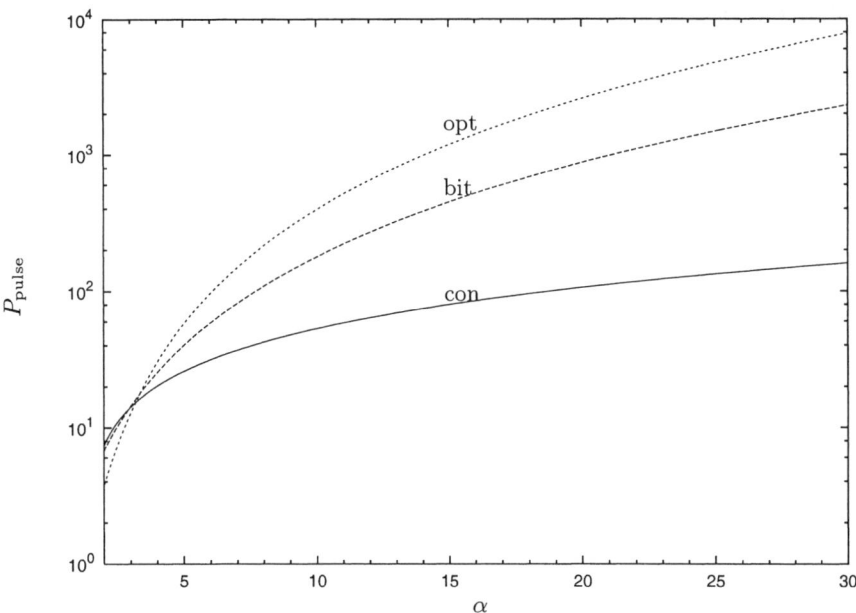

Fig. 2.26. Power P_{pulse} in units of $\frac{\pi}{\mu_0} \frac{\xi}{\lambda^2} \frac{a_1 \sigma_{\max}^2}{\mathcal{F}_{\text{Mat}}(T_i, T_f)}$ as a function of the parameter α for the three coil types: the coil with constant current density ('con'), the Bitter coil ('bit') and the stress-optimized coil ('opt'). Both coil parameters were set equal, $\alpha = \beta$

time, so that the ohmic losses are still small. We find

$$P_{\text{pulse}} = \frac{2W_{\text{m}}}{t_{\text{pulse}}} = \frac{\pi}{\mu_0} \frac{\xi}{\lambda^2} \frac{a_1 \sigma_{\max}^2}{\mathcal{F}_{\text{Mat}}(T_i, T_f)} \mathcal{P}(\alpha, \beta) \quad . \tag{2.116}$$

Again, $\mathcal{P}$ depends only on the coil type and the coil parameters α and β. For the three coil types we find

$$\mathcal{P}_{\text{con}} = \frac{16}{3} \beta \quad (\text{for} \quad 2 \leq \alpha) \,, \tag{2.117}$$

$$\mathcal{P}_{\text{bit}} = \frac{\beta \alpha^2}{(\ln \alpha)^2} \,, \tag{2.118}$$

$$\mathcal{P}_{\text{opt}} = \frac{\beta \alpha^2}{\ln \alpha} \,. \tag{2.119}$$

2.2.6 Examples for Pulsed Coils

We give here examples for coils made with some representative conductor materials. The coils were calculated for a fixed inner radius of 1 cm. For the

Table 2.2. Data for the geometry and the conductor materials used for the calculation of pulsed coils

geometry	a_1	=	0.010	m	inner radius
	a_2		variable		outer radius
	h	=	$2\,a_2$		coil height
	T_i	=	77	K	initial temperature
	T_f	=	400	K	final temperature
	λ	=	0.8		filling factor
copper	ϱ	=	2.8×10^{-9}	$\Omega\,\text{m}$	resistivity
	$\sigma_{\max}$	=	400	MPa	tensile strength
	F_{Mat}	=	9.42×10^{16}	$\text{A}^2\,\text{s}\,\text{m}^{-4}$	material integral
copper-beryllium	ϱ	=	1.5×10^{-8}	$\Omega\,\text{m}$	resistivity
	$\sigma_{\max}$	=	750	MPa	tensile strength
	F_{Mat}	=	4.45×10^{16}	$\text{A}^2\,\text{s}\,\text{m}^{-4}$	material integral
maragin steel	ϱ	=	3.2×10^{-7}	$\Omega\,\text{m}$	resistivity
	$\sigma_{\max}$	=	1700	MPa	tensile strength
	F_{Mat}	=	2.71×10^{15}	$\text{A}^2\,\text{s}\,\text{m}^{-4}$	material integral

temperature increase the interval from 77 K to 400 K was used. Three different materials were taken: hard-drawn copper, copper-beryllium (C17510) and a maragin steel (AerMet 100), [95–98].

Maragin steel [99, 100] is the name of a special class of high strength steels that differ from conventional steels in that they are hardened by a metallurgical reaction that does not involve carbon. Instead, these steels are strengthened by the precipitation of intermetallic compounds at temperatures of about 780 K. The term 'maragin' is derived from 'martensitic age hardening' and denotes age hardening of a low-carbon martensitic matrix. These steels typically have very high nickel, cobalt and molybdenum contents and very low carbon contents. Carbon, in fact, is an impurity in these steels and is kept as low as commercially feasible. The absence of carbon and the use of intermetallic precipitation to achieve hardening produce several unique characteristics that set maragin steels apart from conventional steels. Hardenability is of no concern because the low-carbon martensite formed after annealing is relatively soft. During age hardening there occur only very slight dimensional changes. Thus, fairly intricate shapes can be machined in the soft condition and then hardened with a minimum of distortion. Weldability is excellent and fracture toughness is considerably better than that of conventional high-strength steels. This characteristic in particular has led to the use maragin steels in many demanding applications.

The results for the three materials are shown in Figs. 2.27–2.29. The central field is given as a function of the outer radius a_2. Furthermore, the relation between the central field and the magnetic energy W_m, the necessary power P_{pulse} and the pulse length t_{pulse} is shown. The height h in the expression of the magnetic energy was set to $h = 2\,a_2$, so that W_m is approximately the magnetic energy of a coil with equal height and outer diameter. The pulse length is that of an ideal rectangular current pulse. The necessary power to be delivered to the coil is estimated from $P_{\text{pulse}} = \frac{2W_m}{t_{\text{pulse}}}$, i.e. the source has to deliver the magnetic energy W_m within a time $t_{\text{pulse}}/2$ to the coil. The resistive power is neglected in this estimation.

The general behavior of the graphs is the same for all three materials; they differ only in their numerical values. For a fixed outer radius a_2 or for a fixed amount of magnetic energy W_m the stress-optimized coil generates more field than the coil with constant current density. The field from the latter saturates at a value that depends on the yield strength of the material used.

The relation of field versus power shows a region where the coil with constant current density is more power efficient than the stress-optimized coil. The differences are rather small, however, and regarding the rough estimation for the power may not hold for a real magnet with finite length.

The stress optimization leads to one severe disadvantage, namely a drastic shortage of the possible pulse length t_{pulse}.

2.3 Radial Transmission of Forces

In this chapter we investigate infinitly long coils, where a transmission of forces in the radial direction is allowed. We treat the case of a coil with constant current density and the case of the Bitter coil. Additionally the coil can have an outer reinforcement or lie in an external field. For the mechanical properties of the materials used we assume the isotropic elastic case, i.e the material is characterized by the Young's modulus and the Poisson ratio.

Again we use the trick of extending the real coil with inner radius a_1, outer radius a_2 and height h to infinity in the axial direction, so that we can calculate the fields and stresses within the coil.

We use in this chapter only the boundary condition of finite mechanical stresses. For any coil design the pulse length would result from setting the boundary condition of finite heat capacity equal to that of the finite mechanical strength, as outlined in Sect. 2.2.5.

2.3.1 Coil with Constant Current Density

Coil with Outer Reinforcement. Here the coil has an outer wall, which serves as reinforcement and exerts a pressure p_a on the coil (see Fig. 2.30).

74   2. Analytical Calculations

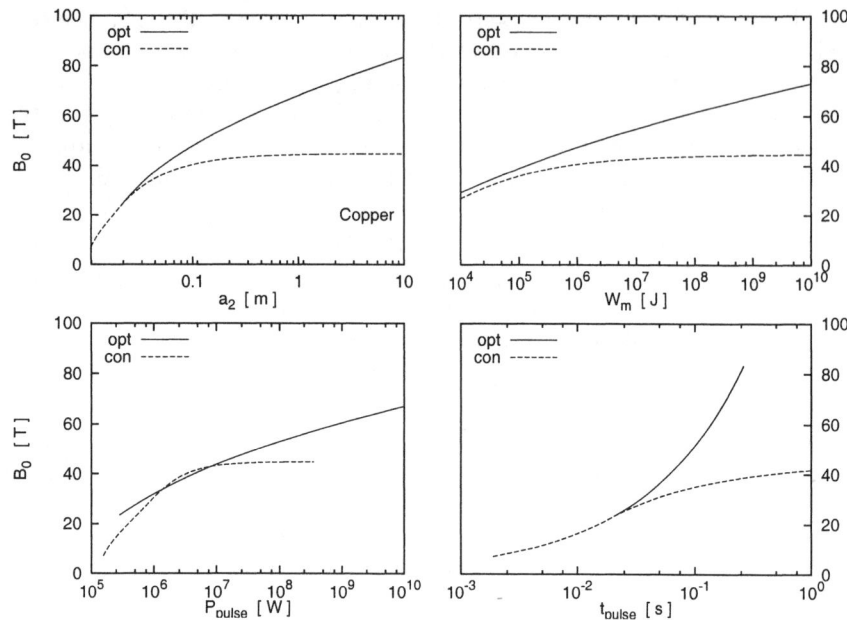

Fig. 2.27. Pulsed coil made from copper. Shown are the relations between the central field B_0 and the outer radius a_2, the magnetic energy W_m, the mean power P_{pulse} and the length of a rectangular current pulse t_{pulse} for a coil with constant current density ('con') and a stress-optimized coil ('opt'). The data used are from Table 2.2

This is a model for a wire-wound coil with outer reinforcement. The equations for the current density and the field are the same as in Sect. 2.2.2. The radial transmission of forces has an influence on the mechanical stress tensor, however. Solving the equilibrium condition $\nabla \boldsymbol{\sigma} + \boldsymbol{f} = 0$, we find now two non-vanishing components of the stress tensor, an azimuthal and a radial one. Again both can be calculated analytically; however, they are quite long expressions. For the energy density of the magnetic field in the center we use the abbreviation

$$p_i := \frac{B_0^2}{2\mu_0} , \qquad (2.120)$$

where B_0 is the field in the center. For a magnet with a very thin wall thickness $a_2 - a_1$ this is the pressure of the magnetic field exerted on the current-carrying wall. We find for the radial and azimuthal component of the stress tensor:

2.3 Radial Transmission of Forces   75

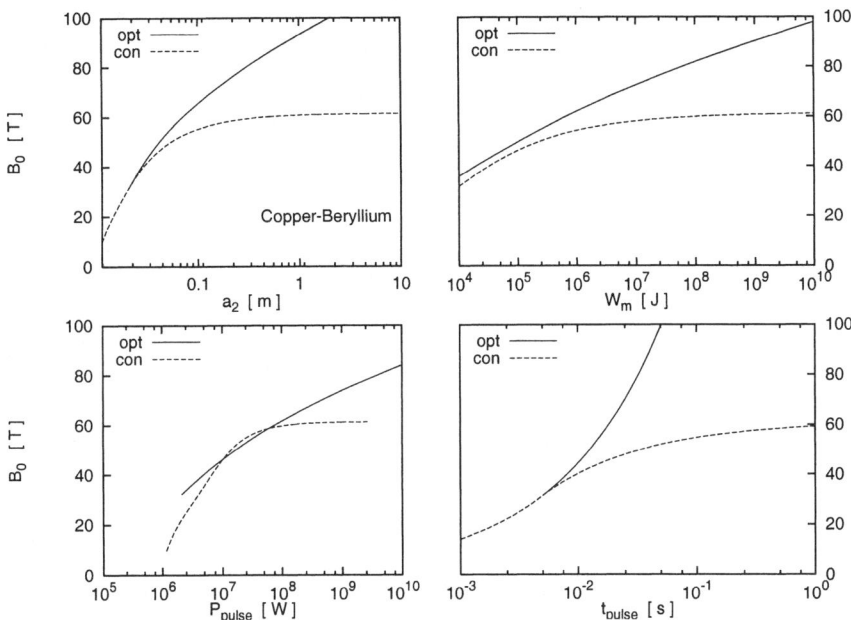

Fig. 2.28. Pulsed copper-beryllium coil (C17510) for the constant current density ('con') and the stress-optimized case ('opt'). The necessary input data are from Table 2.2. The central field B_0 is plotted as a function of the outer radius a_2, the magnetic energy W_m, the power P_{pulse} and the pulse length t_{pulse}

$$\frac{\sigma_{rr}(x)}{p_i} = \left[g(1) - g(\alpha) - \frac{p_a}{p_i}\right] \frac{\alpha^2}{x^2} \frac{x^2 - 1}{\alpha^2 - 1} - [g(1) - g(x)] , \quad (2.121)$$

$$\frac{\sigma_{\varphi\varphi}(x)}{p_i} = \left[g(1) - g(\alpha) - \frac{p_a}{p_i}\right] \frac{\alpha^2}{x^2} \frac{x^2 + 1}{\alpha^2 - 1} - [g(1) - g(x)]$$

$$+ \frac{2(1-\nu)}{(\alpha-1)^2} \left[\frac{\alpha}{3} - \frac{x}{4}\right] x \quad (2.122)$$

with the abbreviation

$$g(x) = \frac{2}{(\alpha-1)^2} \left[\frac{3+\nu}{8} x - \frac{2+\nu}{3} \alpha\right] x . \quad (2.123)$$

Here ν is the Poisson ratio of the conductor material and x is the radial position in units of the inner radius, $x = r/a_1$. For most materials where $\nu \approx 1/3$, the equations may be simplified to

$$g(x) = \frac{2}{(\alpha-1)^2} \left[\frac{5}{12} x - \frac{7}{9} \alpha\right] x . \quad (2.124)$$

With only these two non-zero stress components the von Mises stress is

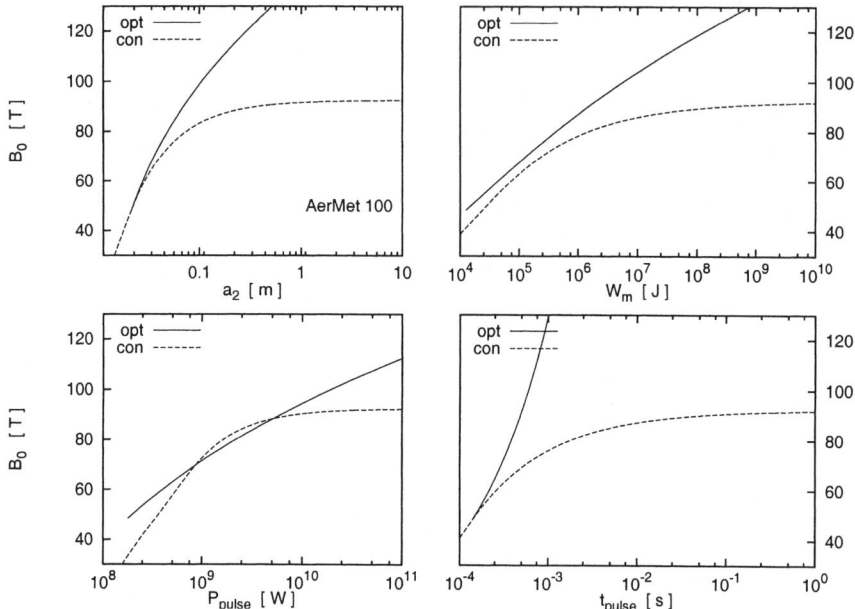

Fig. 2.29. Coil made out of maragin steel (AerMet 100). Shown are the central field B_0 vs a_2, W_m, P_{pulse} and t_{pulse} for a constant current density coil ('con') and a stress-optimized coil ('opt'). The calculations were performed with the data from Table 2.2

$$\sigma_{\text{vonMises}} = \sqrt{\sigma_{rr}^2 - \sigma_{rr}\,\sigma_{\varphi\varphi} + \sigma_{\varphi\varphi}^2}\,. \tag{2.125}$$

The von Mises stress for a coil with $\alpha = 20$ is shown in Fig. 2.31 as a function of the normalized radius $x = r/a_1$ and the normalized pressure p_a/p_i of the outer shell. One can show that at the inner surface $x = 1$ the von Mises stress falls off linearly with increasing outer pressure p_a down to 0, then it increases again linearly. At the outer surface $x = \alpha$ the von Mises stress increases monotonously with increasing outer pressure. Therefore, there exists a certain pressure where the stress at the inner surface is equal to that at the outer surface of the coil. Because the stress values inside the coil are lower, this represents the best situation with the overall lowest levels of stress. The optimal value for the outer pressure p_a is calculated to be

$$\frac{p_a}{p_i} = \frac{17\alpha^2 + 2\alpha + 15}{18(\alpha^2 + 1)}\left\{1 - \sqrt{1 - \frac{32(\alpha^2+1)(7\alpha^2+2\alpha+9)}{(17\alpha^2+2\alpha+15)^2}}\right\}, \tag{2.126}$$

which converges for large α to

$$\frac{p_a}{p_i} = 0.497 \tag{2.127}$$

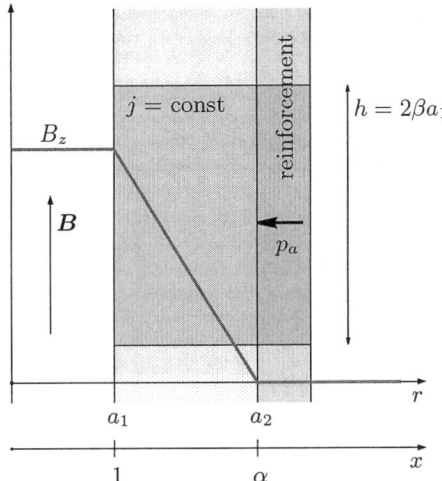

Fig. 2.30. Setup for a coil with constant current density. The mechanical forces are also transmitted in the radial direction, and an outer reinforcement exerts a pressure p_a on the coil. In dimensionless units the coil has an inner radius of 1 and an outer radius of α. The coil has inner radius a_1, outer radius a_2 and height h. The coil with its outer reinforcement is shown in a darker color. For calculating the fields and the stresses we use the trick of extending the coil to infinity in the axial direction

and the maximal von Mises stress therefore is

$$\sigma_{\text{vonMises}} = 0.45\, p_i \;, \tag{2.128}$$

where we used the definition (2.125) of the von Mises stress and (2.121) and (2.122). Resolving for the central field yields (see (2.120))

$$B_0 = 2.11\, \sqrt{\mu_0 \sigma_{\max}} \quad \text{for} \quad p_a = 0.497\, p_i \;. \tag{2.129}$$

Without reinforcement ($p_a = 0$) the von Mises stress would have for large α a maximum value of $\frac{13}{9}\, p_i$ at $x = 1$, and the central field would be

$$B_0 = 1.18\, \sqrt{\mu_0 \sigma_{\max}} \quad \text{for} \quad p_a = 0 \;. \tag{2.130}$$

We conclude that for coils with constant current density and with radial transmission of forces an outer reinforcement is of much benefit.

Coil in an External Field. Here we consider a coil lying in a constant background field B_a. The stress tensor in the coil has a radial and an axial component; these are

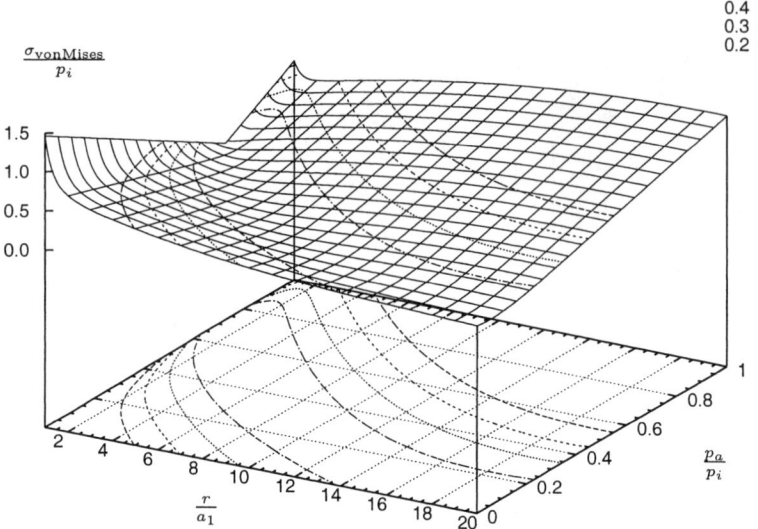

Fig. 2.31. Von Mises stress as a function of the normalized radius r/a_1 and the normalized outer pressure p_a/p_i for a coil with $\alpha = 20$. The stress is given in units of $p_i = \frac{B_0^2}{2\mu_0}$. The overall lowest levels of the von Mises stress are achieved if the stress at the inside of the coil, $x = 1$, is equal to that at the outside, here $x = 20$. This is the case if the outer pressure is about half the inner pressure

$$\frac{\sigma_{rr}(x)}{p_i} = [g(1) - g(\alpha)]\frac{\alpha^2}{x^2}\frac{x^2-1}{\alpha^2-1} - [g(1) - g(x)] \,, \tag{2.131}$$

$$\frac{\sigma_{\varphi\varphi}(x)}{p_i} = [g(1) - g(\alpha)]\frac{\alpha^2}{x^2}\frac{x^2+1}{\alpha^2-1} - [g(1) - g(x)]$$
$$+ \frac{2(1-\nu)}{(\alpha-1)^2}(1-b)\left[\frac{\alpha-b}{3} - \frac{1-b}{4}x\right]x \tag{2.132}$$

where

$$g(x) = \frac{2(1-b)}{(\alpha-1)^2}\left[\frac{3+\nu}{8}(1-b)x - \frac{2+\nu}{3}(\alpha-b)\right]x \tag{2.133}$$

and the magnetic pressure p_i and the ratio of the background field B_a to the the central field B_0 are given by

$$p_i = \frac{B_0^2}{2\mu_0} \quad \text{and} \quad b = \frac{B_a}{B_0}\,. \tag{2.134}$$

In Fig. 2.32 the von Mises stress (see (2.125)) is shown for a coil with $\alpha = 20$ as a function of the normalized outer field B_a/B_0 and the normalized radius $x = r/a_1$. The Poisson ratio was set to $\nu = \frac{1}{3}$. The highest levels of stress

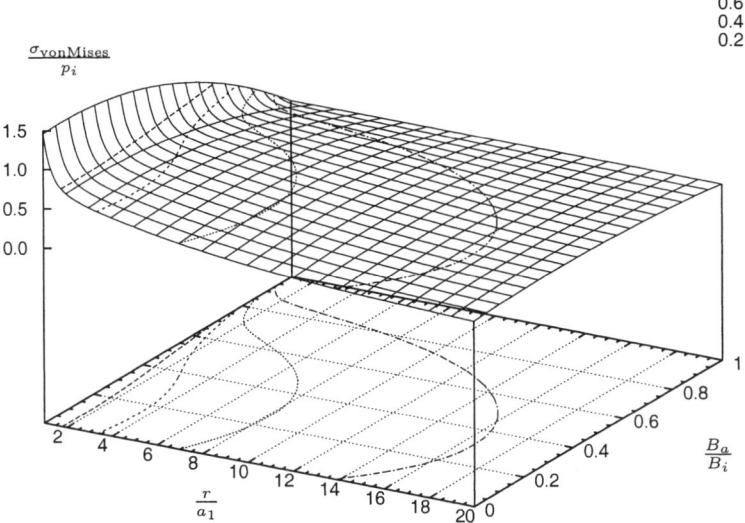

Fig. 2.32. von Mises stress as a function of the normalized radius $x = r/a_1$ and the normalized outer field B_a/B_0 for a coil with $\alpha = 20$. The stress is given in units of p_i. Independent of the outer field the highest levels of stress occur always at the inside of the coil at $x = 1$. This maximum is highest when no outer field is applied; it decreases then with increasing outer field. In order to reduce the peak stress quite high outer fields are necessary, however

occur always at the inside of the coil at $x = 1$, independent of the outer field. However, the height of this maximum decreases with increasing outer field. An outer field of $B_a = 0.6 B_0$, for instance, reduces the peak stress from approximately 1.5 to 1.0 in relative units. This means that for reducing the maximal stresses quite strong external magnetic fields are necessary. For comparison with the coil with the reinforcing cylinder, therefore, a two-coil system has to be investigated.

Two Nested Coils. The setup of the two coils is as follows: the inner coil stretches in the radial direction from a_1 to a_2 and the outer coil from $a_2+\delta$ to a_3, where the gap δ is taken as infinitly small. The maximum possible stresses in the inner and outer coil are σ_1 and σ_2, respectively. The mechanical forces are transmitted in the radial direction, except at the gap at a_2. The current density in each coil is supposed to be constant, but may be different for the inner and outer coil. The material of the coils is assumed to be elastic and isotropic. The magnetic field is constant in the central bore, B_0. Within the inner coil it falls off linearly to B_2 at a_2, then it drops linearly to zero in the outer coil. The situation is sketched in Fig. 2.33. We define

$$\alpha = \frac{a_3}{a_1} \quad \text{and} \quad \alpha_1 = \frac{a_2}{a_1} \,. \tag{2.135}$$

2. Analytical Calculations

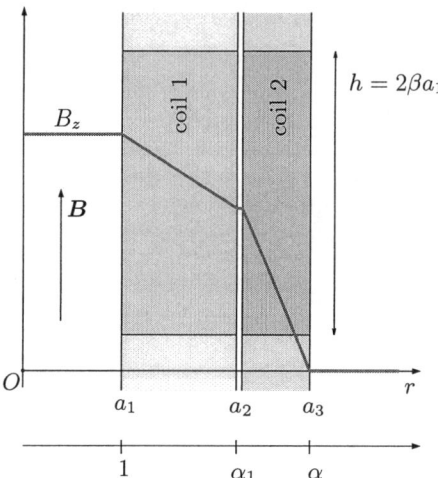

Fig. 2.33. Geometry for the case of two nested coils. The inner coil stretches in radial direction from 1 to α_1 in dimensionless units, the outer coil from $\alpha_1 + \delta$ to α. The gap δ is assumed to be very small. Both coils are assumed to have constant current densities, j_1 and j_2; they may not be equal, however. The same holds for the yield strengths σ_1 and σ_2. The mechanical forces are transmitted in the radial direction, except at the gap at α_1. Both coils have height h. Again, we extend the coils to infinity in the axial direction in order to calculate the fields as well as the stresses

The task is now to find for any given α the best α_1, i.e. the intersection radius which results in the highest central field under the boundary condition that the stress limits of both coils are not exceeded. The calculation of the central field in its dependence on α, α_1, σ_1 and σ_2 is somewhat lengthy but straightforward; it just combines the results from the last section. The field B_2 at the gap a_2 is a function of α, α_1 and σ_2:

$$\frac{B_2}{\sqrt{\mu_0\,\sigma_2}} = \sqrt{\frac{18\,(\alpha^2 - \alpha_1^2)}{13\alpha^2 + 2\alpha\alpha_1 + 3\alpha_1^2}}\,. \qquad (2.136)$$

The central field depends on α, α_1, σ_1 and, via B_2, also on σ_2:

$$\frac{B_0}{\sqrt{\mu_0\,\sigma_1}} = -\frac{\alpha_1^2 - 1}{13\alpha_1^2 + 2\alpha_1 + 3}\frac{B_2}{\sqrt{\mu_0\,\sigma_1}}$$

$$+ \sqrt{\left(\frac{14\alpha_1^2 + 2\alpha_1 + 2}{13\alpha_1^2 + 2\alpha_1 + 3}\right)^2 \frac{B_2^2}{\mu_0\,\sigma_1} + \frac{18\,(\alpha_1^2 - 1)}{13\alpha_1^2 + 2\alpha_1 + 3}}\,. \qquad (2.137)$$

For any given set of α, σ_1 and σ_2 there exists one α_1 which results in the highest central field. A graph of the central field as a function of α and α_1 is shown in Fig. 2.34 for the special case of $\sigma_1 = \sigma_2$.

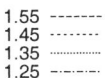
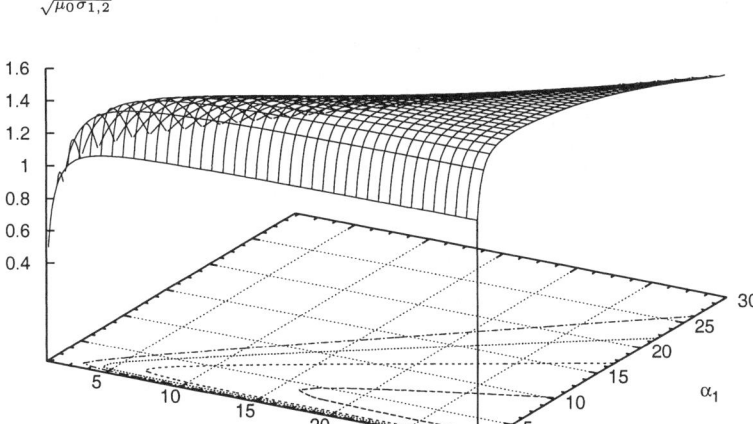

Fig. 2.34. Central field B_0 in units of $\sqrt{\mu_0 \sigma_1}$ as function of the relative outer radius α and the intersection radius α_1, which separates the inner from the outer coil. The maximal stresses were set equal here, $\sigma_1 = \sigma_2$

For $\alpha = 20$, for instance, the maximum lies at about $\alpha_1 = 4.3$ and has a value of $B_0 = 1.56 \sqrt{\mu_0 \sigma_{1,2}}$. The corresponding field at the gap a_2 is $B_2 = 1.12 \sqrt{\mu_0 \sigma_{1,2}}$.

2.3.2 Bitter Coil

Coil with Outer Reinforcement. Here we investigate an infinitly long coil with a current density proportional to $1/r$, i.e. a Bitter coil. Allowing for a radial transmission of the mechanical forces we find for the radial and azimuthal component of the stress tensor:

$$\frac{\sigma_{rr}(x)}{p_i} = g(x) - \frac{\alpha^2}{x^2} \frac{x^2 - 1}{\alpha^2 - 1} \left[\frac{p_a}{p_i} + g(\alpha) \right] , \qquad (2.138)$$

$$\frac{\sigma_{\varphi\varphi}(x)}{p_i} = g(x) - \frac{\alpha^2}{x^2} \frac{x^2 + 1}{\alpha^2 - 1} \left[\frac{p_a}{p_i} + g(\alpha) \right]$$

$$+ \frac{1 - \nu}{2 (\ln \alpha)^2} [1 + 2\ln \alpha - 2\ln x] , \qquad (2.139)$$

where

$$g(x) = [(1 - \nu) - (1 + \nu)(2\ln \alpha - \ln x)] \frac{\ln x}{2 (\ln \alpha)^2} , \qquad (2.140)$$

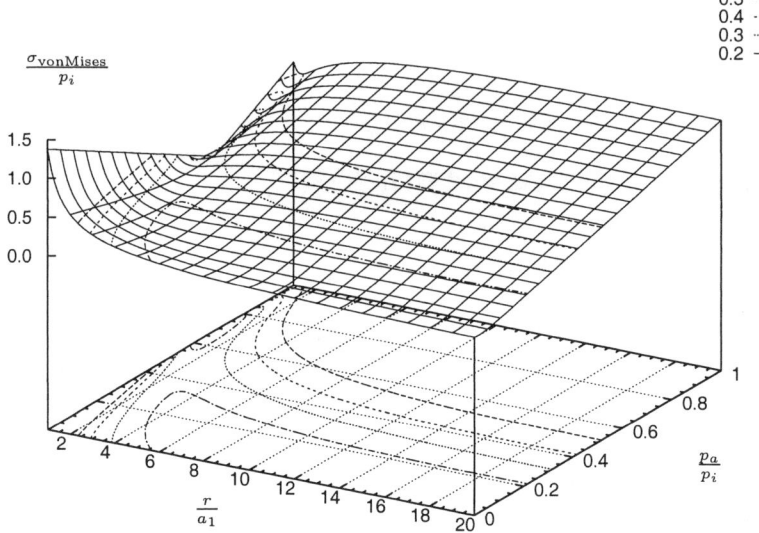

Fig. 2.35. Von Mises stress as a function of the normalized radius $x = r/a_1$ and the normalized outer pressure p_a/p_i for a Bitter coil with $\alpha = 20$. The stress is given in units of $p_i = \frac{B_0^2}{2\mu_0}$. The overall lowest levels of the von Mises stress are achieved if the stress at the inside of the coil, $x = 1$, is equal to that at the outside, here $x = 20$. This is the case if the outer pressure is about half the inner pressure

where p_i is the magnetic pressure in the center and p_a is the pressure of the reinforcement onto the coil.

In Fig. 2.35 the von Mises stress is shown for a Bitter coil with $\alpha = 20$ as a function of the normalized radius $x = r/a_1$ and the outer pressure p_a. As in the case of the coil with constant current density, there exists an optimal value of the outer pressure p_a which makes the von Mises stress in the coil minimal. The optimal value for the outer pressure p_a is calculated to be (we use $\nu = 1/3$)

$$\frac{p_a}{p_i} = \frac{4(4\alpha^2 + 3)(\ln \alpha)^2 + 3(\alpha^2 - 1 - 2\ln \alpha)}{18(\alpha^2 + 1)(\ln \alpha)^2} \qquad (2.141)$$

$$\times \left\{ 1 - \sqrt{1 - 48(\alpha^2 + 1)(\ln \alpha)^2 \frac{4(\alpha^2 + 1)(\ln \alpha)^2 + 2(\alpha^2 - 1 - 2\ln \alpha)}{[4(4\alpha^2 + 3)(\ln \alpha)^2 + 3(\alpha^2 - 1 - 2\ln \alpha)]^2}} \right\},$$

which converges for large α to

$$p_a = 0.444\, p_i, \qquad (2.142)$$

and the maximum von Mises stress therefore becomes

$$\sigma_{\text{vonMises}} = 0.444\, p_i, \qquad (2.143)$$

where we have used the definition (2.125) of the von Mises stress and (2.121) and (2.122). With (2.120) we can resolve for the central field yields and find:

$$B_0 = 2.12\sqrt{\mu_0 \sigma_{\max}} \quad \text{for} \quad p_a = 0.444\, p_i \ . \tag{2.144}$$

For a Bitter coil with $\alpha = 20$ the optimal outer pressure is calculated as $p_a = 0.463\, p_i$. The maximum von Mises stress and the central field turn out to be

$$p_a = 0 \quad \sigma_{\text{vonMises}} = 1.373\, p_i \quad B_0 = 1.21\sqrt{\mu_0 \sigma_{\max}}, \tag{2.145}$$

$$p_a = 0.463\, p_i \quad \sigma_{\text{vonMises}} = 0.445\, p_i \quad B_0 = 2.12\sqrt{\mu_0 \sigma_{\max}} \ . \tag{2.146}$$

Again we can conclude that for massive coils with transmission of the mechanical forces in the radial direction an outer reinforcing shell is of much benefit in reducing the von Mises stress in the winding area of the coil, which in turn has an influence on the maximum achievable central field.

Coil in an External Field. Now we put the Bitter coil into a constant background field, B_a, which is for convenience expressed in units of the central field:

$$B_a = b\, B_0 \ . \tag{2.147}$$

The two non-vanishing components of the stress tensor now become:

$$\frac{\sigma_{rr}(x)}{p_i} = g(x) - \frac{\alpha^2}{x^2} \frac{x^2 - 1}{\alpha^2 - 1} g(\alpha) \ , \tag{2.148}$$

$$\frac{\sigma_{\varphi\varphi}(x)}{p_i} = g(x) - \frac{\alpha^2}{x^2} \frac{x^2 + 1}{\alpha^2 - 1} g(\alpha)$$

$$\times \frac{1 - \nu}{2} \left(\frac{1 - b}{\ln \alpha}\right)^2 \left[1 + \frac{2 \ln \alpha}{1 - b} - 2 \ln x\right] \ , \tag{2.149}$$

where

$$g(x) = \left[(1-\nu) - (1+\nu)\left(\frac{2\ln\alpha}{1-b} - \ln x\right)\right] \left[\frac{1-b}{\ln\alpha}\right]^2 \frac{\ln x}{2} \ . \tag{2.150}$$

In Fig. 2.36 the von Mises stress (see (2.125)) is shown for a Bitter coil with $\alpha = 20$ as a function of the normalized outer field B_a/B_0 and the normalized radius $x = r/a_1$. As was the case with the coil with constant current density in a background field, the highest levels of stress occur always at the inside of the coil at $x = 1$. This maximum is highest when no outer field is applied; it then decreases with increasing outer field. Considerable background fields are necessary in order to reduce the peak stresses at the inner bore, however.

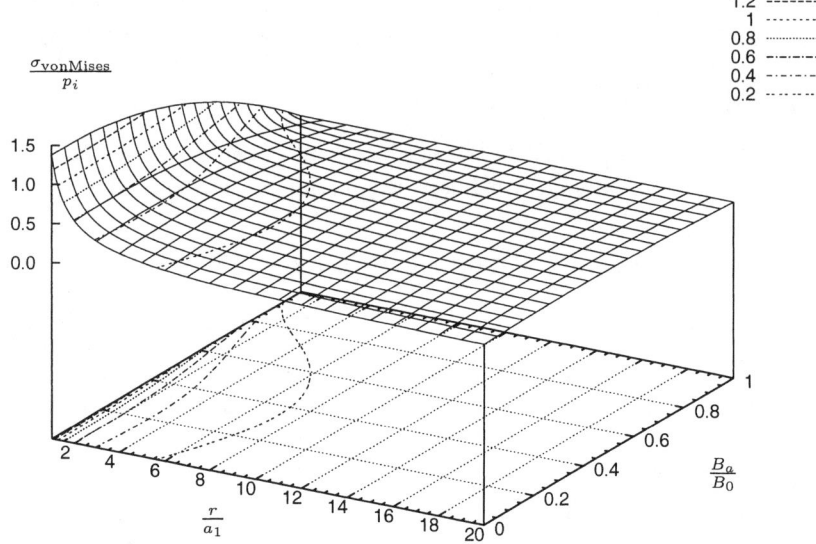

Fig. 2.36. Von Mises stress as a function of the normalized radius $x = r/a_1$ and the external field B_a for a Bitter coil with $\alpha = 20$. The stress is given in units of $p_i = \frac{B_0^2}{2\mu_0}$

Conclusion. For an overview we summarize here the results for the various coil designs from Sects. 2.2 and 2.3. For infinitely long coils with constant current density we investigated four different coil types:

1. the polyhelix coil, where no transmission of forces in radial direction occurs, (2.101),
2. a coil with radial transmission of forces, but without reinforcement, Sect. 2.3.1 with $p_a = 0$,
3. a coil with radial transmission of forces with the reinforcement exerting the ideal pressure, Sect. 2.3.1,
4. two nested coils, each with the same constant current density and radial transmission of forces, Sect. 2.3.1.

The central field of these coil types as a function of the coil parameter α is shown in Fig. 2.37. In the case of $\alpha = 20$ we get for the central field of the four coil types with constant current density:

$$\frac{B_0}{\sqrt{\mu_0 \sigma_{\max}}} = \begin{cases} 1.90 \text{ no radial force transmission, polyhelix} \\ 1.17 \text{ radial force transmission, no reinforcement} \\ 2.10 \text{ radial force transm., reinforcement, } p_a = 0.502\, p_i \\ 1.56 \text{ radial force transmission, two nested coils} \end{cases}.$$

In the case of the two nested coils we have assumed here the same current density and yield strength for the inner and outer coil. Because of the constant

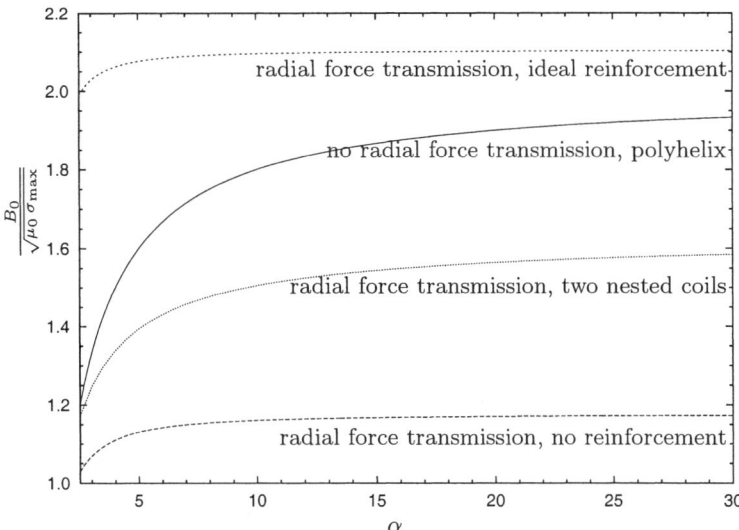

Fig. 2.37. Normalized central field as a function of the outer coil radius α for several coil types with constant current density. In the parameter range shown the best coil type is the one with radial transmission of forces in combination with an outer reinforcement. This coil is followed by the polyhelix coil and the two-coil system. The least best design is the coil with radial force transmission without any reinforcement

current density and the same α, all the above coils have the same values of necessary power and stored magnetic energy. Using (2.68–2.71) we find ($\alpha > 2$ assumed) for the magnetic energy W_m and the power P_{pulse} associated to a coil of length 2α:

$$W_m = \pi\,\sigma_{\max}\,a_1^3\,2\alpha\,\frac{\alpha^2+2\alpha+3}{3}\left(\frac{\alpha-1}{\alpha}\right)^2,$$

$$P_{\text{pulse}} = \frac{\pi}{\mu_0}\,\frac{\xi}{\lambda^2}\,\frac{a_1\,\sigma_{\max}^2}{\mathcal{F}_{\text{Mat}}(T_i,T_f)}\,\frac{16}{3}\,\frac{(\alpha^2+2\alpha+3)(\alpha-1)^2}{\alpha^3}.$$

The best technique for a coil with constant current density would therefore be to allow for radial transmission of forces, combined with an outer reinforcing shell. The follow-up design would be the polyhelix design, which allows no transmission of mechanical forces in the radial direction. The two-coil technique is the first step from a pure coil with radial force transmission, the least best design, towards a polyhelix. Making not only two subcoils, but three, four, etc., would make this coil technique converge towards the polyhelix design.

In the case of Bitter coils we have dealt with three coil techniques:

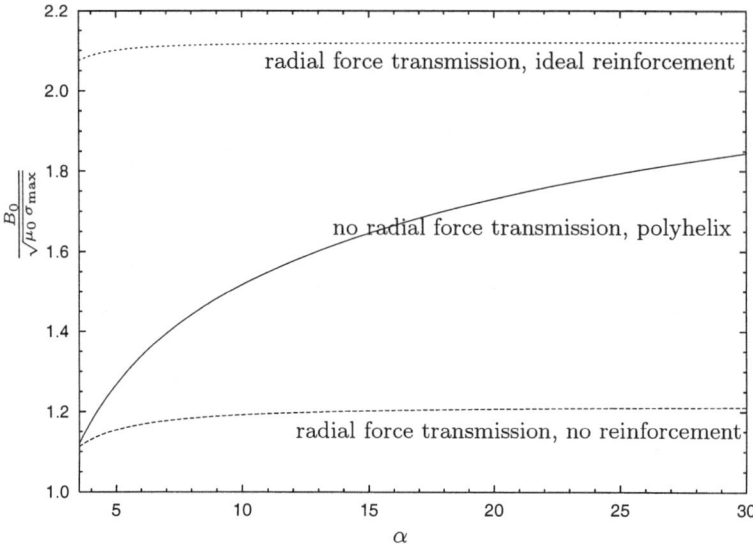

Fig. 2.38. Normalized central field as a function of the outer coil radius α for several types of Bitter coil. In the parameter range shown the best coil type is the one with radial transmission of forces in combination with an outer reinforcement. A coil without any reinforcement is the least best design; the polyhelix design lies between these two coil techniques

1. the polyhelix coil, in which no transmission of forces in the radial direction occurs, (2.102),
2. a coil with radial transmission of forces, but without reinforcement, Sect. 2.3.2 with $p_a = 0$,
3. a coil with radial transmission of forces with the reinforcement exerting the ideal pressure, Sect. 2.3.2.

The central fields for the three coil types are shown in Fig. 2.38 as a function of the outer coil radius α. For the special case of $\alpha = 20$ we find

$$\frac{B_0}{\sqrt{\mu_0 \, \sigma_{\max}}} = \begin{cases} 1.73 & \text{no radial force transmission} \\ 1.21 & \text{radial force transmission, no reinforcement} \\ 2.12 & \text{radial force trans., reinforcement, } p_a = 0.463 \, p_i \end{cases}$$

And finally for a stress-optimized coil with $\alpha = 20$ (see (2.103)) we have

$$\frac{B_0}{\sqrt{\mu_0 \, \sigma_{\max}}} = 2.45 \quad \text{(stress-optimized polyhelix, no radial force transmission)}.$$

Table 2.3 compares the results for the various coil types for the special case of $\alpha = 20$. Shown are the central fields, the magnetic energy in the coil (we assume a coil of axial length 2α), the pulse time and the estima-

Table 2.3. Summary of the coils from Sects. 2.2 and 2.3. Shown are the magnetic energy W_m in the coil, the pulse length t_{pulse}, the estimation of the necessary power P_{pulse} and the central field B_0 for various coil techniques. All quantities are given in relative units

	W_m	t_{pulse}	P_{pulse}	B_0
constant current density	5.33×10^3	100	107	
no radial force transmission, polyhelix				1.90
radial force transmission, no reinforcement				1.17
radial force transmission, reinforcement, $p_a = 0.502\, p_i$				2.10
radial force transmission, two nested coils				1.56
Bitter coil	1.31×10^3	3	876	
no radial force transmission				1.73
radial force transmission, no reinforcement				1.21
radial force transmission, reinforcement, $p_a = 0.463\, p_i$				2.12
stress optimized coil	8.10×10^3	6	2698	
stress-optimized polyhelix coil, no radial force transmission				2.45

tion of the necessary power. All quantities are given in relative units, i.e.

the central field B_0 in units of $\sqrt{\mu_0 \sigma_{\max}}$,

the magnetic energy W_m in units of $\pi \sigma_{\max} a_1^3$,

the pulse length t_{pulse} in units of $(\mu_0 \lambda a_1)^2 \dfrac{\mathcal{F}_{\text{Mat}}(T_i, T_f)}{\xi \mu_0 \sigma_{\max}}$, and

the power P_{pulse} in units of $\dfrac{\pi}{\mu_0} \dfrac{\xi}{\lambda^2} \dfrac{a_1 \sigma_{\max}^2}{\mathcal{F}_{\text{Mat}}(T_i, T_f)}$.

Table 2.3 reveals also that each coil technique has its advantages and disadvantages. The stress-optimized coil gives the highest possible fields, the Bitter coil needs the least amount of magnetic energy and the coil with constant current density can deliver the longest pulse durations.

3. Numerical Simulations

3.1 Polyhelix Coils

This section is the numerical follow-up to Sects. 2.1 and 2.2. We consider a rectangular coil of the polyhelix type [90–94]. The aim of the numerical optimization is to find the distribution of the current density which yields the highest possible field in the center of the coil. The optimization is performed under the constraints of the maximum allowed von Mises stress in the coil and a finite amount of magnetic energy.

For the special coil model used in Sect. 2.2, where the coil is thought to extend to infinity in the axial direction, we know already that the optimal coil has a constant von Mises stress throughout the coil winding and the necessary current density takes on a valley-like shape (see Fig. 2.20). The trick with the infinite extension in the axial direction allowed us to calculate fields and stresses by analytical methods. For real coils with finite length this is no longer possible and one has to resort to numerical methods.

The real polyhelix considered in this chapter incorporates the axial as well as the radial component of the magnetic field, and uses the hoop stress $\sigma_{\varphi\varphi}$ and the axial stress σ_{zz}. The radial field components and axial stresses were both neglected in Sect. 2.2. The technical realization of a polyhelix coil might look like Fig. 2.10.

3.1.1 Coil Model

The numerical simulation requires us to fully specify the coil model. We assume here a polyhelix coil with rectangular cross-section (see Fig. 3.1). The current density distribution is taken as continuous as a function of only the radial position, $j = j(r)$. This ideal polyhelix is for numerical purposes approximated by a finite number N of helices; in the following simulation we use $N = 20$. Each of these helices has its own (constant) current density, which allows us to approximate the current density $j(r)$ with a stepwise constant distribution. The magnetic field and the vector potential from each helix is calculated by the methods outlined in Sect. 1.1.7. We use (1.64–1.66) for the field calculations and (1.52) to get the vector potential of a helix. A summation of all helices gives the fields and the vector potential for the polyhelix.

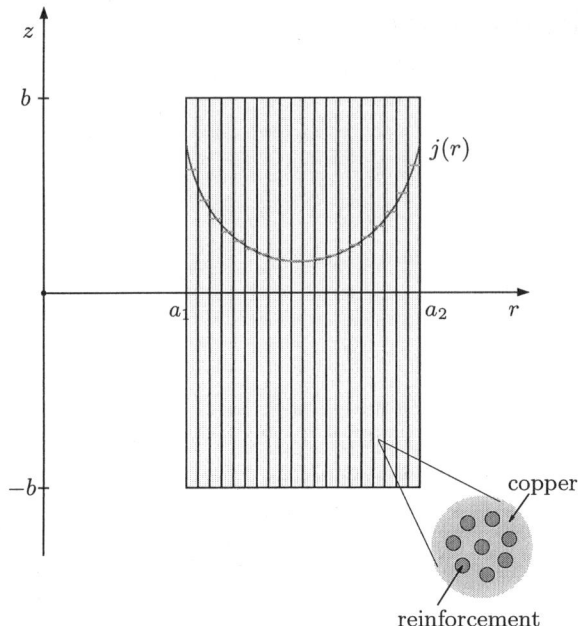

Fig. 3.1. Polyhelix coil with inner radius a_1, outer radius a_2 and height $2b$, the current density in the polyhelix is a function of the radial position only, $j = j(r)$. For the simulation this ideal polyhelix is approximated by $N = 20$ helices, each having the same thickness $(a_2 - a_1)/N$ and height $2b$. The current density in each helix is assumed to be constant; the continuous current density $j(r)$ of the ideal polyhelix is hence approximated by a stepwise constant current density function. Between each helix there will be no mechanical transmission of forces, so that one can still assume the decoupling of the equilibrium condition $\nabla \boldsymbol{\sigma} + \boldsymbol{f} = 0$ to be valid. The finer structure of the distribution of the stresses in each helix is neglected here. The material properties are assumed to be constant and isotropic; on a microscopic scale we think of a composite, consisting of copper as conductor and an unspecified reinforcement. The yield strength of the composite follows from a simple mixing rule of the two ingredients and their respective share in the cross-section. Furthermore, the current will flow only through the copper, which determines the heating properties and resistivity of the composite

The inductance matrix of the N helices is calculated via (1.32). This $N \times N$-matrix is calculated once at the beginning of the simulation, which then allows us to calculate the magnetic energy of any current density distribution $j(r)$ in the coil very efficiently from

$$W_{\text{mag}} = \frac{1}{2} \sum_{i,k=1}^{N} L_{ik} I_i I_k \, , \tag{3.1}$$

where I_i and I_k denote the current in the helix i and k, respectively.

3.1 Polyhelix Coils

Lorentz forces in the radial as well as the axial direction are taken into account. The mechanical stresses in the coil are calculated for the case of an idealized polyhelix coil, i.e. there is no transmission of mechanical forces in the radial direction. This permits a very fast calculation of the mechanical stresses in the coil, because in the equilibrium condition (see Sect. 1.2)

$$\nabla \boldsymbol{\sigma} + \boldsymbol{f} = 0 \tag{3.2}$$

the radial and the azimuthal equations become decoupled in this case ($\boldsymbol{\sigma}$ denotes the stress tensor and $\boldsymbol{f}$ the density of the Lorentz forces):

$$-\frac{1}{r}\sigma_{\varphi\varphi} + f_r = 0 , \tag{3.3}$$

$$\partial_z \sigma_{zz} + f_z = 0 , \tag{3.4}$$

and the radial and axial components of the Lorentz force density are

$$f_r = j\, B_z \quad \text{and} \quad f_z = -j\, B_r . \tag{3.5}$$

The stress tensor has two components: the hoop stress $\sigma_{\varphi\varphi}$ and the axial stress σ_{zz}. Now, we assumed the current density to be only a function of the radial position, which allows us to integrate (3.4) and express it by the vector potential at the point (r, z) and the point (r, b) at the topside of the coil (half-height $= b$). Hence, the two stress components are

$$\sigma_{\varphi\varphi}(r,z) = j(r)\, r\, B_z(r,z) , \tag{3.6}$$

$$\sigma_{zz}(r,z) = j(r) \left[A_\varphi(r,b) - A_\varphi(r,z) \right] . \tag{3.7}$$

For the model of an ideal polyhelix coil the solution of the equilibrium condition can be expressed as a function of the axial field, the current density and the vector potential. This is the reason why the stress calculation for a polyhelix is so fast, for the solution of the coupled equilibrium condition is generally a major computational task.

With only these two components of the stress tensor, the hoop stress $\sigma_{\varphi\varphi}$ and the axial stress σ_{zz}, the von Mises stress becomes

$$\sigma_{\text{vonMises}} = \sqrt{\sigma_{\varphi\varphi}^2 - \sigma_{\varphi\varphi}\sigma_{zz} + \sigma_{zz}^2} . \tag{3.8}$$

For symmetry reasons the highest von Mises stress will always occur in the midplane of the coil. For any algorithm determining the maximum von Mises stress it is therefore sufficient to search only in the midplane of the coil for the highest stresses.

For the simulation with $N = 20$ helices we also assume the decoupling of the equilibrium condition to be valid. The stress within one helix is furthermore assumed to be constant. In a real helix there would be a variation of the stresses, because there is a transmission of the mechanical forces in the radial direction. This means that there is a radial stress component within each helix and also that the equilibrium condition in each helix is no longer decoupled, which would again make any numerical calculation difficult.

The material properties (yield strength, specific heat, specific resistivity) are assumed to be homogeneous and isotropic, and the coil will be operated in the elastic regime. As the material for the wire we use a microcomposite consisting of copper as the conductor and a reinforcement. The filling factor of this composite will be $\lambda = 1$, and we neglect any insulation or cooling channels or other voids. The yield strength $Y_{\text{composite}}$ of the composite follows by a mixing rule from both materials and their respective share in the cross-section:

$$Y_{\text{composite}} = (1 - x)\, Y_{\text{conductor}} + x\, Y_{\text{reinforcement}}, \qquad (3.9)$$

where $(1 - x)$ parts are covered by the conductor and x parts by the reinforcement.

The current is assumed to flow only through the copper, and any heat transfer into the reinforcement is neglected. This means that the ratio of copper to reinforcement determines the amount of adiabatically heated copper. The relation between initial temperature, final temperature, current density and pulse length is given by (1.97) and is expressed here as

$$j^2_{\text{conductor}}\, t_{\text{pulse}}\, \xi = \mathcal{F}_{\text{conductor}}(T_{\text{i}}, T_{\text{f}}), \qquad (3.10)$$

or, if we use the properties of the composite, as

$$j^2_{\text{composite}}\, t_{\text{pulse}}\, \xi = \mathcal{F}_{\text{composite}}(T_{\text{i}}, T_{\text{f}}). \qquad (3.11)$$

For a rectangular current pulse we have $\xi = 1$. The pulse length of the polyhelix coil is determined by the helix with the highest current density $j_{\text{composite}}$.

3.1.2 Optimization with Constraints

The numerical problem to be solved can be defined as follows. Find for a coil of the above-mentioned type and with a given cross-section the optimal current density distribution $j(r)$ which gives the highest field in the center under the following three conditions:

Condition 1: The von Mises stress should stay below some given value.

Condition 2: There is only a certain amount of magnetic energy available.

Condition 3: There is an upper limit for the current density.

With the first condition we can assure that the coil always stays in its elastic regime. For that purpose we require the maximum von Mises stress to be lower than the yield strength of the coil material.

The second boundary condition is necessary to restrict the size of the coil. Without that condition a search algorithm would not converge, for the generated field of a certain coil could be increased by increasing the size of the coil and still comply with the constraint of a finite mechanical stress. This process would never stop, we would find that we can produce infinitly high fields with an infinite amount of magnetic energy. We have encountered that fact already with the stress-optimized coil in Sect. 2.2.5.

The third condition ensures that there exists a lower boundary for the pulse length t_{pulse} (see (1.97)), since

$$t_{\text{pulse}} = \frac{\mathcal{F}_{\text{composite}}(T_i, T_f)}{\xi\, j^2_{\text{max,composite}}} , \qquad (3.12)$$

which is defined by the chosen coil material (material integral $\mathcal{F}_{\text{Mat}}(T_i, T_f)$ from (1.95) and the initial and final temperatures T_i and T_f, respectively. The factor ξ reflects the shape of the current pulse through the coil: $\xi = 1$ holds for a rectangular current pulse and $\xi = 1/2$ for a half of a sine wave. The maximum current density of all the helices is denoted by $j_{\text{max,composite}}$.

Mathematically speaking, this is an NLP problem with constraints (NLP stands for **N**onlinear **P**rogramming). There are numerous algorithms and techniques available for handling such problems; an overview can be found in [101–103], and details about certain algorithms may be found in [104–112]. The following calculations were performed with the code 'SolvOpt' (**Solv**er for local **Opt**imization problems), which is concerned with the minimization or maximization of nonlinear problems and accounts for constraints by the method of exact penalization [113].

3.1.3 The Optimal Current Density

We apply now the 'SolvOpt' algorithm to a polyhelix coil with an inner radius of 10 mm, outer radius of 200 mm and a height of 400 mm. This total cross-section is divided into 20 helices, and the current densities in these helices are optimized by the algorithm in order to achieve the highest possible field in the center. The boundary conditions are an upper current density of $1.0 \times 10^9 \text{A m}^{-2}$, a maximal von Mises stress of 1 GPa in the coil and an available magnetic energy of 50 MJ. The material of the coil is a composite consisting of copper with a yield strength of 320 MPa and a reinforcement of 2 GPa, and the ratio of both components is adjusted to reach the desired yield strength of the composite of 1 GPa. The coil is subjected to a rectangular current pulse, the length of which is determined by the initial temperature of 77 K and the final temperature of 400 K, in combination with the coil material.

The results of the optimization with the 'SolvOpt' algorithm are shown in Table 3.1. For comparison with the stress-optimized coil the data of a reference coil with constant current density are also shown. The current density

Table 3.1. Coil data of Sect. 3.1.3: definition of the coil geometry and admissible temperature range, the composition of the composite material and the used boundary conditions in the SolvOpt algorithm. The lower part of the table shows the results of the simulation. In order to estimate the quality of the solution, the optimized coil is compared with a reference coil, which has a constant distribution of the current density in all 20 helices and also complies with the three boundary conditions. Graphs of the solution can be found in Fig. 3.2

Coil data:			
inner radius	0.010	m	
outer radius	0.200	m	
coil height	0.400	m	
initial temperature	77	K	
final temperature	400	K	
number of helices	20		
Composite:			
nominal yield strength	1.0	GPa	
conductor	0.32	GPa	copper
reinforcement	2.0	GPa	
Boundary conditions:			
maximum von Mises stress	1.0	GPa	condition 1
maximum magnetic energy	50	MJ	condition 2
maximum current density	1.0×10^9	$A\,m^{-2}$	condition 3

Simulation results:			
	optimized	reference	coil
central field	77.6	60.1	T
highest current density	9.9	2.9	$10^8\,A\,m^{-2}$
pulse length	34	403	ms
magnetic energy	19.5	12.4	MJ
thermal energy	3.5	29.3	MJ
initial time constant	1.17	1.63	s
final time constant	0.18	0.15	s

in this reference coil was adjusted to generate a maximum von Mises stress of 1 GPa. The optimized coil generates much more field (77.6 T versus 60.1 T), but needs much more magnetic energy (15.5 MJ versus 12.4 MJ) and has a shorter pulse length (34 ms versus 403 ms). In Fig. 3.2 the field and stresses in the midplane of the polyhelix as well as the current density and the final temperature are shown. The axial field in the midplane decreases from the

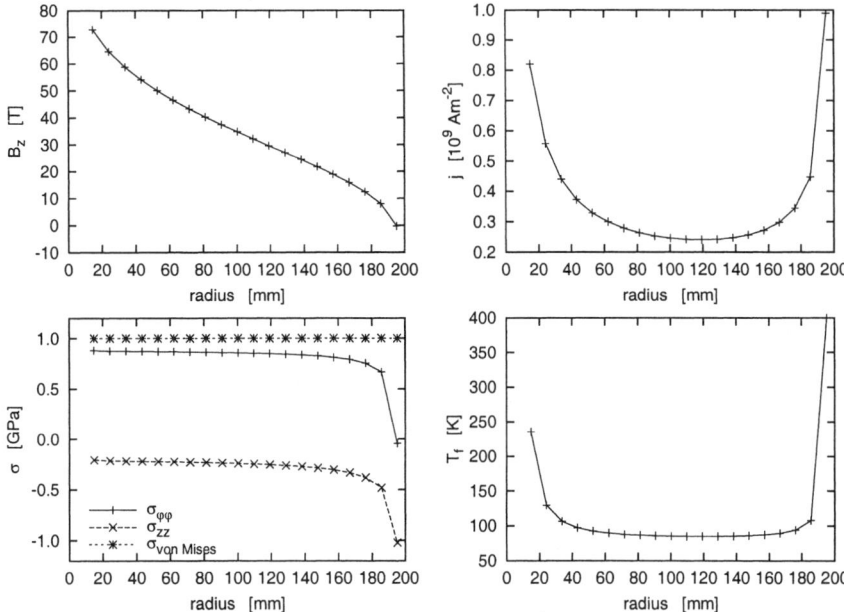

Fig. 3.2. Simulation results for the polyhelix of Sect. 3.1.3 with the SolvOpt algorithm. The four figures show the axial field B_z and the stresses in the midplane (von Mises stress σ_{vonMises}, hoop stress $\sigma_{\varphi\varphi}$ and axial stress σ_{zz}), the current density j and the final temperature T_f in the polyhelix coil. Compare the functions with those from Fig. 2.20, where radial fields and axial stresses were neglected!

innermost helix towards the outermost helix; the von Mises stress along the same path remains constant at the maximum admissible value of 1 GPa. The current density of the stress-optimized coil forms a 'valley': it has highest values for the inner and outer helix and is lower for all other helices. This is the same behavior as encountered in Sect. 2.2.4 (see especially Fig. 2.20) for the model of a stress-optimized coil where we neglected the radial field components and the axial stress components.

The highest current density, occurring in the outermost helix, determines the pulse length of 34 ms. During that time the rectangular current pulse heats the outermost helix to the desired final temperature of 400 K. Because the current density in all the other helices is much lower, this also means that the final temperatures are also lower. The distribution of the final temperatures in the helices follows the same pattern as the current densities, as can clearly be seen from Fig. 3.2. This explains the much lower value for the thermal energy for the stress-optimized coil with respect to the reference coil (3.5 MJ versus 29.3 MJ), because in the latter coil every helix becomes heated up to the final temperature of 400 K.

3.1.4 Variation of the Stored Magnetic Energy

As a second example we optimize one and the same coil for several different values of the magnetic energy stored in the coil. The dimensions of the coil, the chosen maximum stress capability of the conductor material and the maximum current density are given in Table 3.2.

Table 3.2. Coil from Sect. 3.1.4: definition of the coil geometric data and admissible temperature range, the composition of the composite material and the used boundary conditions in the SolvOpt algorithm. The lower part of the table shows the central fields for simulations where the coil was subjected to different values of magnetic energy. The stress constraint limits the magnetic energy in this coil to 19.5 MJ; the much higher available energy in the last two calculations cannot be used. A few selected graphs of the current density distribution in the polyhelix are shown in Fig. 3.3

Coil data:					
inner radius	0.010	m			
outer radius	0.200	m			
coil height	0.400	m			
initial temperature	77	K			
final temperature	400	K			
number of helices	20				
Composite:					
nominal yield strength	1.0	GPa			
conductor	0.32	GPa	copper		
reinforcement	2.0	GPa			
Boundary conditions:					
maximum von Mises stress	1.0	GPa	condition 1		
maximum current density	1.0×10^9	$A\,m^{-2}$	condition 3		
Simulation results:					
W_{mag} [MJ]	B_0 [T]	t_p [ms]	W_{mag} [MJ]	B_0 [T]	t_p [ms]
1.0	52.4	33	12.0	75.0	46
2.0	60.1	33	13.0	75.4	47
3.0	65.2	34	14.0	75.9	47
4.0	67.5	37	15.0	76.3	48
5.0	69.2	39	16.0	76.6	48
6.0	70.5	40	17.0	76.9	49
7.0	71.5	42	18.0	77.2	49
8.0	72.4	43	19.0	77.5	46
9.0	73.2	44	19.5	77.6	34	(20 MJ)
10.0	73.9	45	19.5	77.6	34	(25 MJ)
11.0	74.5	45			

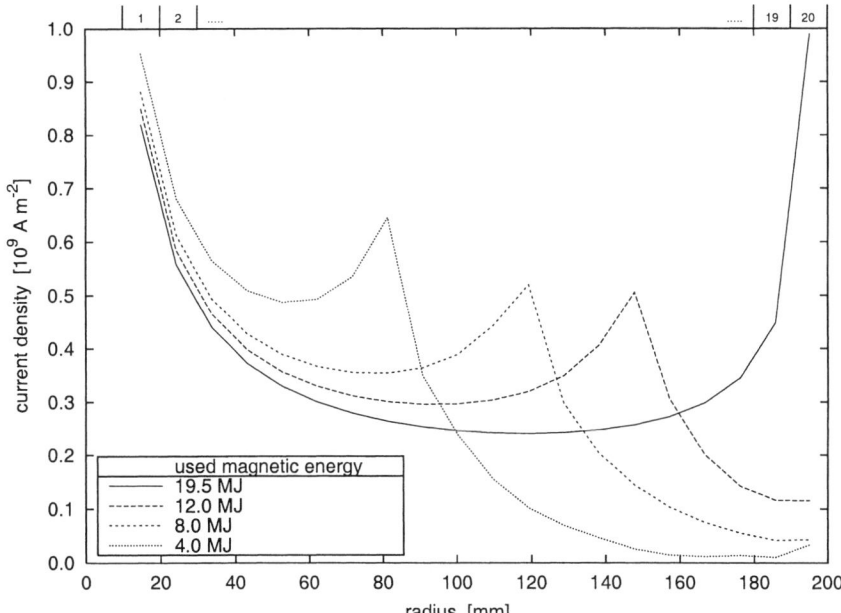

Fig. 3.3. Current density distribution for some of the data from Table 3.2. The current density has a peak at the inner surface of the coil, then it drops to form a valley with a second peak near the outer radius of the coil. Then $j(r)$ drops rapidly. The location of this second peak depends on the amount of magnetic energy in use. The higher this value, the more the second peak moves outwards, until it reaches the outer radius, which in this example is the case for a magnetic energy in the coil of 19.5 MJ. For this amount of magnetic energy the coil is used at its limits of stress capability and current density. Subjecting this coil to more magnetic energy would violate one of the boundary conditions (stress or current density), and with less magnetic energy the coil would be over-dimensioned. The graph to 4.0 MJ shows, for instance, that some of the outer helices carry almost no current and could very well be omitted

For this coil the SolvOpt algorithm then determines the optimal current density distribution $j(r)$ resulting in the highest field in the center. As an additional constraint we limit the available magnetic energy to a certain value. The calculations were performed with magnetic energies varying from 1 MJ up to 25 MJ.

The results are shown in Table 3.2, which lists the achieved central magnetic field as a function of the magnetic energy in the coil. The constraint of a maximum von Mises stress of 1 GPa allows this coil to use only a maximum magnetic energy of 19.5 MJ; the allowed 20 or 25 MJ in two of the calculations cannot be used.

Graphs of the current density are shown in Fig. 3.3. For reasons of clarity, only a subset of the calculations of Table 3.2 is drawn. All graphs of the current density have a peak at the inner surface of the coil, then they drop to form a valley with a second peak near the outer radius of the coil, and finally $j(r)$ drops rapidly. The location of this second peak depends on the amount of magnetic energy in use. The higher this value is, the more the peak moves radially outwards, until it reaches the outer radius, which in our coil here is the case for 19.5 MJ. With this amount of magnetic energy the coil is used at its limits of stress capability and current density.

With less energy the coil is not used at its full potential. The second, outer peak in the current density distribution and the fall-off thereafter indicate that the coil is too big; the part beyond the second peak could as well be left out. One could – rather should – make the coil smaller.

The conclusion of these calculations is that the boundary condition of finite available magnetic energy can be a limiting factor for the coil and its optimized current density distribution. If one wants to use a coil at its full potential regarding the stress limit and the upper limit of the current density, then there exists an upper limit of the magnetic energy. The upper limit of the current density is equivalent to a lower limit of the pulse length.

This upper limit of the magnetic energy is also found with the SolvOpt algorithm. We start with an arbitrary boundary value for the admissible magnetic energy and let the SolvOpt algorithm do its work. If the magnetic energy resulting from the simulation is less than the admissible magnetic energy and when the outer peak of the current density distribution lies at the outer radius of the coil, then we have found the maximum admissible magnetic energy. In the examples this is the case for the last two calculations, where we allowed 20 MJ and 25 MJ of magnetic energy, but could use only 19.5 MJ of magnetic energy (see Table 3.2).

3.1.5 Variation of Coil Size and Material Strength

In this series of calculations we vary the outer radius of the coil and the strength of the conductor material. The height of the coil is always taken as equal to its outer diameter. The limit of admissible magnetic energy is taken as sufficiently high – see the last paragraph – so that the physical boundary conditions of the coil are composed of the finite mechanical strength of the composite material and the upper limit of the current density. The coil parameters are summarized in Table 3.3. The outer radius was varied from 75 mm to 300 mm, as yield strengths of the composite coil material of 1.00, 1.25, 1.50, 1.75 and 2.00 GPa were used. The yield strength of the reinforcement was assumed to be 2.0 and 2.7 GPa, respectively; the 2.7 GPa value was used for the composite with yields of 1.75 and 2.00 GPa.

In Fig. 3.4 the results of the calculations are shown. The upper graph gives the central field as a function of the used magnetic energy in the coil, with the strength of the composite material as a parameter. For the same amount of magnetic energy coils made from stronger materials produce more field. All curves show the same tendency: they increase strongly for low values of the magnetic energy and then they flatten out.

Table 3.3. Definition of the coil geometric data and admissible temperature range, the composition of the composite material and the boundary conditions used in the SolvOpt algorithm for the third series of examples, where the outer radius and the yield strength of the composite material are varied (see Sect. 3.1.5)

Coil data:		
inner radius	0.010 m	
outer radius	a_2	from 75 to 300 mm
coil height	$2b$	$b = a_2$
initial temperature	77 K	
final temperature	400 K	
number of helices	20	
Composite:		
nominal yield strength		from 1.0 to 2.0 GPa
conductor	0.32 GPa	copper
reinforcement		2.0 or 2.7 GPa
Boundary conditions:		
maximum von Mises stress	from 1.0 to 2.0 GPa	condition 1
maximum current density	$10^9 \, \text{A m}^{-2}$	condition 3

For the designer of a magnet system this means that one should avoid the 'flat' region, because the magnetic energy is used there in a very inefficient way. For a given strength there exists a useful size, field and magnetic energy. The designer of the coil is responsible for giving a good definition of 'useful' here. Furthermore, one can trade off the yield strength of the material against the magnetic energy. The same field can be generated either by a strong material with a few megajoules or by a weak wire with many megajoules of magnetic energy.

The lower graph of Fig. 3.4 shows the thermal energy Q_{th}, given in units of the magnetic energy, as a function of the magnetic energy. By thermal energy we mean the energy which heats the copper in the coil (adiabatic process) from the initial temperature of 77 K to the final temperature of 400 K. For very small magnetic energies, and hence also small coils, this ratio lies around or even above 1. As the magnetic energy increases the ratio falls rapidly and for magnetic energies above 10 MJ it lies below ≈ 0.2. In terms of energy, stress-optimized coils are determined by the thermal energy and large coils by the magnetic energy.

The small ratio of thermal to magnetic energy for stress-optimized coils is one of their main disadvantages, however. For instance, the coil with a yield strength of 1.0 GPa would probably be operated with magnetic energies around 10 MJ (see Fig. 3.4), which would mean a thermal energy of 2 MJ. For the operation of the coil one would then have to deliver about 10 MJ of magnetic energy into the coil, and then actively have to extract 8 MJ from

100   3. Numerical Simulations

the coil and dump this energy somewhere else. Dumping this energy into the coil itself would lead to serious overheating and finally destruction of the coil.

For a first rough estimation of the necessary total energy for a magnet system one can add the values of the magnetic energy and the thermal energy. This is an overestimation, because one can allow at least a part of the magnetic energy to be dissipated into the coil itself.

3.1.6 Coil with Smaller Inner Radius

Here we present a series of calculations similar to the previous ones. We investigate variations of the outer radius and the material yield strength. The main difference is that the inner radius is reduced here to $a_1 = 5$ mm. For the same amount of magnetic energy and yield strength of the coil material the coil with the smaller inner radius can generate considerably higher fields.

The fixed coil parameters are shown in Table 3.4. As outer radius values between 50 and 200 mm were chosen, the height of the respective coil was always set to its diameter, $a_2 = b$. The yield strength of the composite material ranged from 1.00 GPa to 2.25 GPa in steps of 0.25 GPa. The yield strength of the reinforcement was set to 2.7 GPa.

In Fig. 3.5 the results of the calculations are shown. The upper graph gives the central field as a function of the magnetic energy, with the strength of the conductor material as a parameter. The lower graph of Fig. 3.5 shows the thermal energy Q_{th}, given in units of the magnetic energy, as a function of the magnetic energy.

The general behavior of these graphs is the same as those from the previous example. Because of the smaller inner bore, however, one can produce higher central fields for a given material strength and magnetic energy.

3.1.7 Comparison with Analytical Methods

In this section we compare the numerical simulations for stress-optimized coils with the analytical estimations from Sect. 2.2.4. As a further reference coil we also present the results for a numerically calculated coil with constant current density distribution.

In all three cases we use a polyhelix coil with an inner radius of 10 mm. The composite coil material has a yield strength of 1 GPa, the conductor of this composite is copper and the reinforcement has a yield strength of 2 GPa. The outer radius is varied between 75 and 300 mm, and the height of the coil is always taken equal to its diameter. The coil data are summarized in Table 3.5, and in Fig. 3.6 the central field and the magnetic energy are shown as a function of the outer radius of the coil.

In the investigated range for the outer radius the field generated by the numerically calculated polyhelix coil with constant current density saturates at a level of 61.5 T, while the necessary magnetic energy increases with the

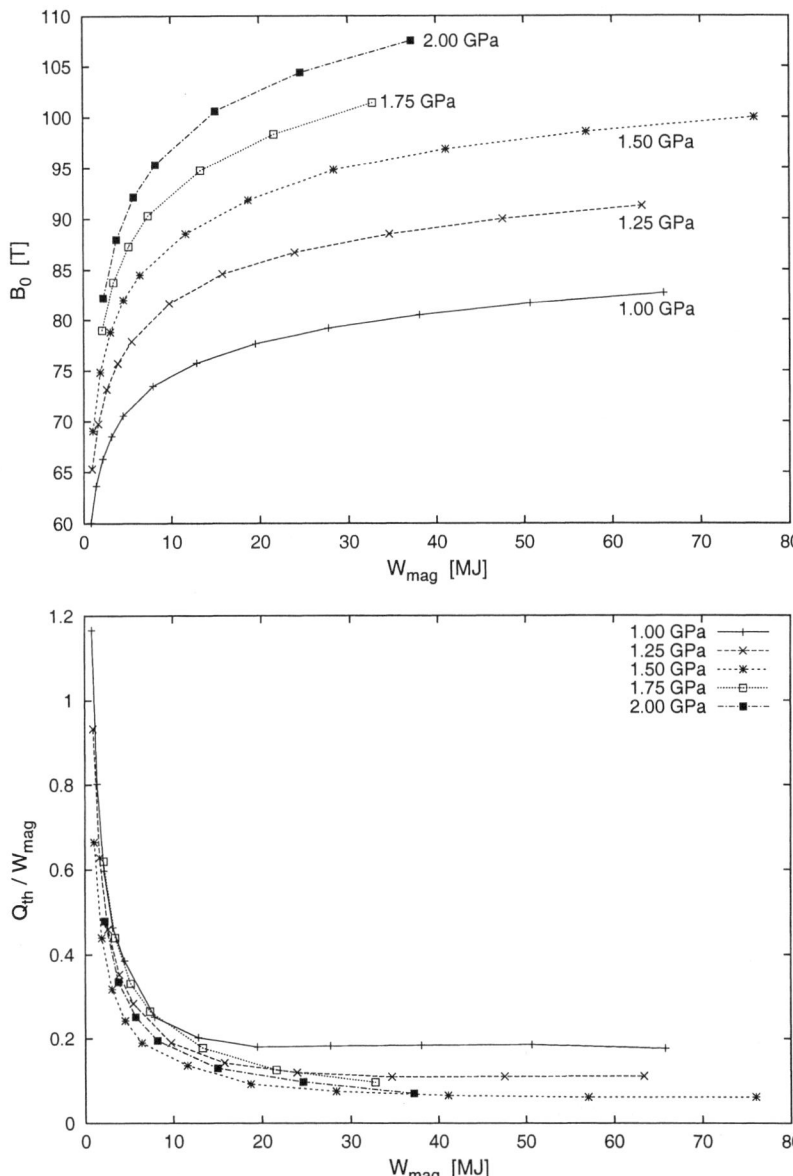

Fig. 3.4. Optimization of coils with an inner radius of 10 mm and various outer radii and yield strengths. The boundary conditions were the indicated yield strengths and an upper limit for the current density; the data can be found in Table 3.3. The upper graph shows the central field, and the lower one the thermal energy, as a function of the magnetic energy. For further details see the text

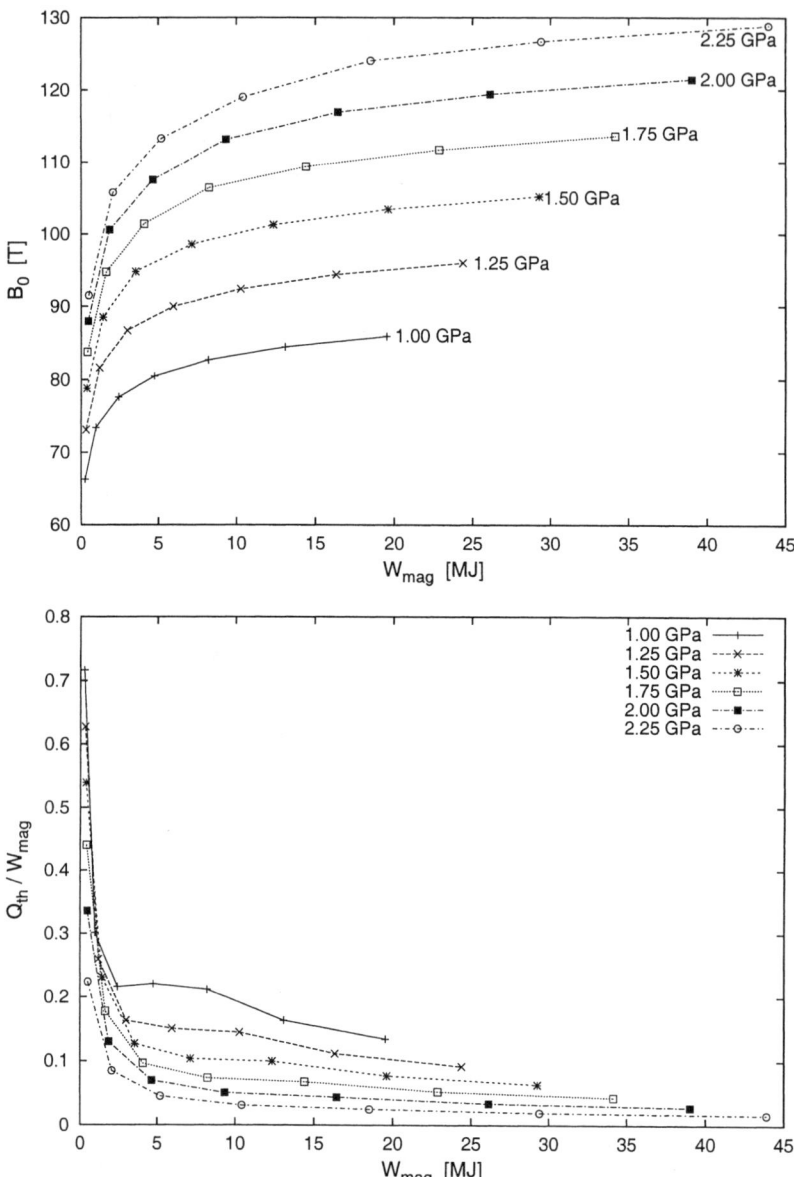

Fig. 3.5. Optimization of polyhelix coils with an inner radius of 5 mm and various outer radii and yield strengths of the composite material. The upper graph shows the central field, and the lower one the thermal energy, as a function of the magnetic energy. For further details see the text and Table 3.4

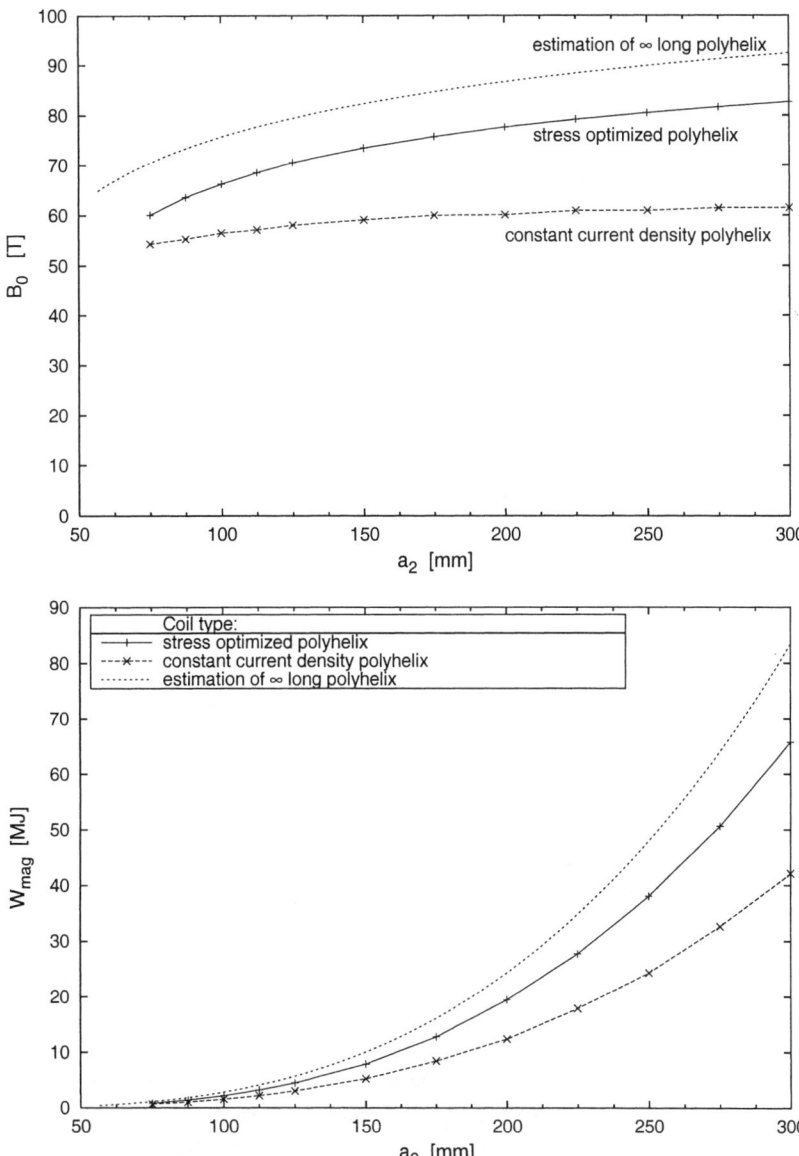

Fig. 3.6. Central magnetic field and magnetic energy as a function of the outer radius a_2 for three different calculations. These are a stress-optimized polyhelix coil and a polyhelix with constant current density. Both of these calculations were performed with numerical methods, and they incorporated axial and radial fields as well as axial stresses and hoop stresses. The third calculation is an estimation for a stress-optimized polyhelix, which was described in Sect. 2.2.4. In all cases the coil data were kept identical (see Table 3.5)

104   3. Numerical Simulations

Table 3.4. Definition of the coil geometric data and admissible temperature range, the composition of the composite material and the boundary conditions used in the SolvOpt algorithm. This data set belongs to Sect. 3.1.6. A series of simulations was carried out for different outer radii and different yield strengths of the composite material of the polyhelix coil. Graphs of the solutions can be found in Fig. 3.4

Coil data:		
inner radius	0.005 m	
outer radius	a_2	from 50 to 200 mm
coil height	$2b$	$b = a_2$
initial temperature	77 K	
final temperature	400 K	
number of helices	20	
Composite:		
nominal yield strength		from 1.00 to 2.25 GPa
conductor	0.32 GPa	copper
reinforcement	2.7 GPa	
Boundary conditions:		
maximum von Mises stress	from 1.00 to 2.25 GPa	condition 1
maximum current density	$1.0 \times 10^9 \, \text{A m}^{-2}$	condition 3

outer radius. Relative to this coil the stress-optimized polyhelix produces much more field at about 1.5 times more magnetic energy. Furthermore, the generated field shows no saturation with increasing outer radius.

Finally, the field and magnetic energy of the estimation of the stress-optimized polyhelix from Sect. 2.2.4 are also shown. This analytical estimation neglected the radial field components and the axial stresses. The graphs of the estimation and the numerical calculation of the stress-optimized polyhelix show the same behavior, as can clearly be seen in Fig. 3.6. However, the central field and the magnetic energy are overestimated by the analytical method. For the field the estimation gives values about 10% too high and for the magnetic energy about 25% too high. Keeping these numbers in mind the estimation from Sect. 2.2.4 is a very useful tool for designing any polyhelix coil.

3.1.8 Estimations and Reality

For the calculations performed here several assumptions have been made. First of all, the coil was treated as an ideal polyhelix. This allowed for a very fast calculation of the mechanical stresses in the coil, because in the corresponding differential equation system, i.e. the equilibrium condition

Table 3.5. Definition of the coil geometric data and admissible temperature range, the composition of the composite material and the boundary conditions used in the SolvOpt algorithm for the comparison of different coil types, where the outer radius was varied (see Sect. 3.1.7)

Coil data:		
inner radius	0.010 m	
outer radius	a_2	from 75 to 300 mm
coil height	$2b$	$b = a_2$
initial temperature	77 K	
final temperature	400 K	
number of helices	20	
Composite:		
nominal yield strength	1.0 GPa	
conductor	0.32 GPa	copper
reinforcement	2.0 GPa	
Boundary conditions:		
maximum von Mises stress	1.0 GPa	condition 1
maximum current density	$10^9\,\mathrm{A\,m^{-2}}$	condition 3

$$\nabla \boldsymbol{\sigma} + \boldsymbol{f} = 0 \,, \tag{3.13}$$

the radial and the azimuthal equations become decoupled ($\boldsymbol{\sigma}$ denotes the stress tensor and $\boldsymbol{f}$ the density of the Lorentz forces). This permits a fast computation of the stresses, and hence an efficient numerical search algorithm for the optimized coil could be implemented. In contrast, in a real polyhelix only the single helices are mechanically separated, and within each one helix there also exists a radial stress component. The incorporation of this radial stress makes the equilibrium condition again a coupled differential equation system. The solution of such a system is in itself a major computational task, and hence a fast search algorithm for finding the optimal current distribution is almost impossible.

The coil is operated in the elastic regime. This is a very conservative assumption, a real coil would allow for plastic deformation. In this respect the calculated coils have some safety margin built in.

The material properties in a real coil are not isotropic; instead the coil is separated into conductor, insulation and reinforcement. Additionally, the material properties of these components are a function of temperature, field and history (plastic deformation!). As long as the equilibrium condition remains decoupled, the anisotropy of the mechanical material properties (different yield strengths in azimuthal and axial direction for instance) is no issue here.

The division into different material regions causes a finer structure of the stress distribution in the coil, but would have no influence on the average stress distribution. Neglecting the temperature and therefore the time dependence of the material properties is a major simplification. For instance, it is known that the yield strength of steel decreases from 1600 MPa at 77 K to 900 MPa at 300 K (ASTM Type 630). The problem is caused here by the change, not the actual value of the yield strength.

Several factors with an influence on the pulse length were also neglected. One was the assumed filling factor of $\lambda = 1$. Incorporating insulation and other voids, for instance, would cause the filling factor to be smaller than 1, which would reduce the possible pulse length. The same effect would be caused by magnetoresistance and eddy currents. They would cause additional heating and, since the final temperature is fixed, would also reduce the pulse length.

An exact calculation of the necessary energy of a capacitor bank would sum together the magnetic energy at the field maximum, the thermal energy in the coil at this time and additional losses in the current leads and the surrounding circuit.

All in all, the aforementioned effects have no principal influence on the overall design of a polyhelix. They constitute 'second-order' effects, which determine the exact behavior of a polyhelix. In that sense they are important to know, but are not among the foremost design criteria. Of course, the last statement has to be taken with a grain of salt! It depends, for instance, on which type of coil one is talking about. For a coil with very short pulse lengths of the order of microseconds one cannot neglect eddy currents as 'second-order' effects; they are a major issue for such a coil.

3.2 Wire-Wound Coils

This chapter is the numerical equivalent to Sect. 2.3, where we calculated wire-wound coils which allowed a transmission of the mechanical forces in the radial direction. In order to use an analytical approach we used the trick with the infinite extension of the coil in the axial direction. This allowed us then to determine the axial field component B_z and the two non-vanishing components of the stress tensor, the hoop stress $\sigma_{\varphi\varphi}$ and the radial stress σ_{rr}, in analytical form. The radial field component B_r as well as the axial stress σ_{zz} and the shear stress σ_{rz} were neglected. Because of the rotational symmetry of the coil, the components $\sigma_{r\varphi}$ and $\sigma_{z\varphi}$ always disappear. In this chapter we include these formerly neglected components. The equilibrium condition (see (1.81))

$$\nabla \boldsymbol{\sigma} + \boldsymbol{f} = 0 \tag{3.14}$$

is now a coupled differential equation system, which generally can be solved only by numerical methods. One widely used technique is the finite element

Table 3.6. Data of the coil from Sect. 3.2.1, which serves as model for a wire-wound coil. These data form the necessary input for the FEM code [119]

inner radius	10	mm
outer radius	100	mm
height	200	mm
current density	4.03×10^{-8}	$A\,m^{-2}$
Young's modulus	135	GPa
Poisson ratio	0.34	

method (FEM, [114–118]). In the following sections we present a few examples of calculations with such an FEM code [119]. The first example demonstrates the results in a somewhat more extensive way in order to give the reader an overall impression of the advantages and disadvantages of an FEM calculation. In the second example we perform a 'search' for the optimal coil under some given boundary conditions, and finally we show a calculation for a two-coil system. The chapter ends with a comparison of the results of the numerical finite element method and the analytical estimation from Sect. 2.3.

3.2.1 Calculation of a Coil

Here we present the results of a calculation for a model of a wire-wound coil. The coil has a rectangular cross-section, the geometrical dimensions are shown in Table 3.6. The current density flowing through the cross-section is assumed to be constant, and its value is adjusted to generate a maximum von Mises stress in the coil of 1 GPa. Throughout the coil the material is homogeneous and isotropic, so that it can be characterized with only two numbers: the Young's modulus and the Poisson ratio. Furthermore, the coil is assumed to operate in the elastic regime. The finer structure of the cross-section, for instance the wire, insulation, etc., is neglected.

The coil from Table 3.6 generates a central field of 39.5 T. The von Mises stress was by construction adapted to reach a maximum value of 1 GPa which is the reason for the strange value for the current density, and the magnetic energy associated with the coil is calculated to be 0.76 MJ. In Figs. 3.7–3.10 the axial and radial field components B_z and B_r as well as the azimuthal component A_φ of the vector potential are shown.

The field around the center of the coil ($r = z = 0$ in cylindrical coordinates) has the shape of a saddle: it decreases in the axial direction and increases in the radial direction. The increase is rather small, however, and one has to magnify the central region, as is done in Fig. 3.8. The radial field is antisymmetric with respect to the midplane of the coil ($z = 0$); it disappears in the midplane and also on the z-axis. The maxima of the radial field component occur at the upper and the lower ends of the coil. The component A_φ of the vector potential is shown in Fig. 3.10, it has a maximum in the

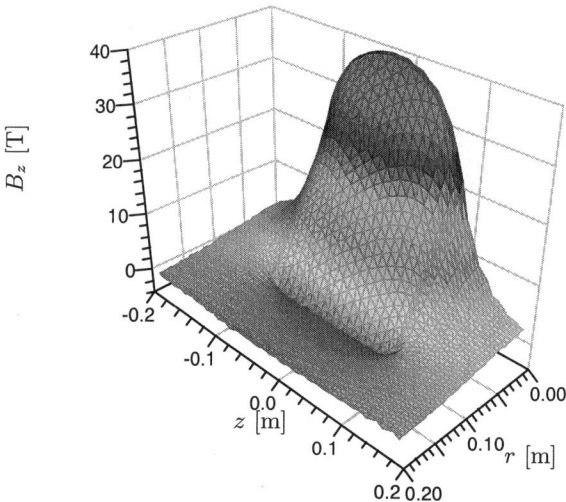

Fig. 3.7. Axial field component B_z from the coil of Table 3.6 as a function of of the radial and axial positions

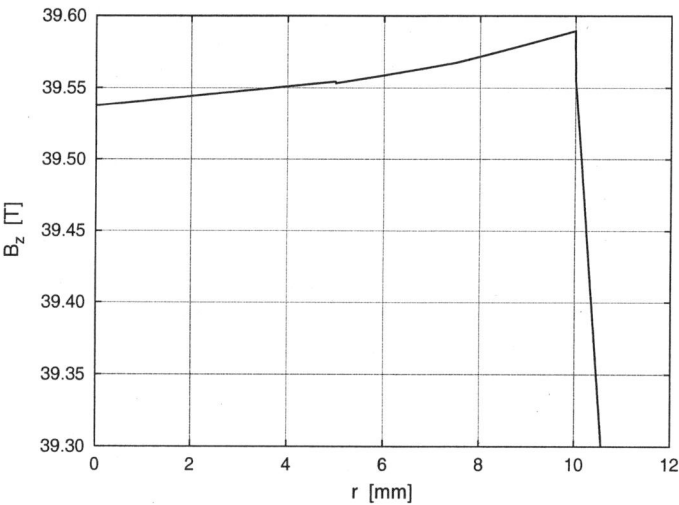

Fig. 3.8. Detail of Fig. 3.7, showing the axial field B_z in the midplane of the coil ($z = 0$) as a function of the radial position r. The inner radius of the coil lies at $r = 0.01$ m; at this point the field has its maximum value. Towards the z-axis of the coil the axial field decreases slightly

midplane of the coil somewhere within the cross-section; along the z-axis it disappears.

The von Mises stress and the four components of the stress tensor, the hoop stress $\sigma_{\varphi\varphi}$, the radial stress σ_{rr}, the axial stress σ_{zz} and the shear

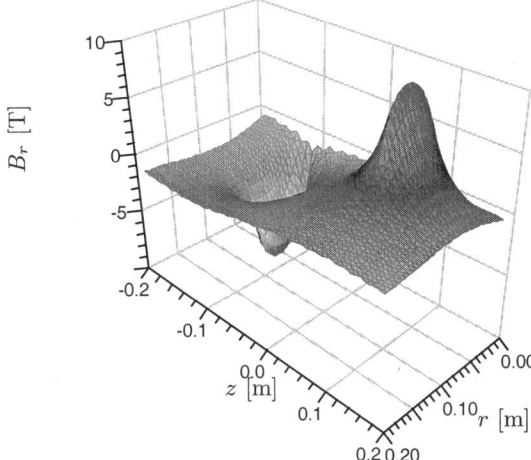

Fig. 3.9. The radial field component B_r from the coil of Table 3.6 as a function of the radial and axial positions. The B_r component is zero on the z-axis and in the midplane of the coil

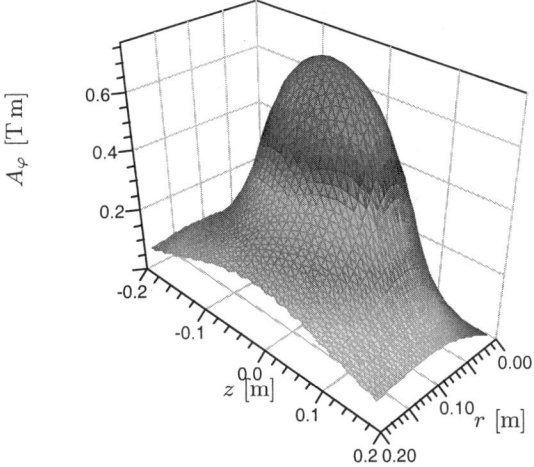

Fig. 3.10. Component A_φ of the vector potential. The maximum of A_φ lies in the midplane at a point within the cross-section of the coil. On the z-axis the vector potential disappears

stress σ_{rz}, are shown in Figs. 3.11–3.15. One should note that the graphs for the various stress components each have a different scaling: for σ_{vonMises} and $\sigma_{\varphi\varphi}$ the range is 0 to 1 GPa, for σ_{rr} it is −20 to 260 MPa, for σ_{zz} it is −140 to 70 MPa and finally for σ_{rz} it is −14 to 14 MPa. Bearing this in mind we see that the hoop stress $\sigma_{\varphi\varphi}$ is the most dominant component of the stress tensor. The next strongest component consists of the radial stress σ_{rr}, followed by

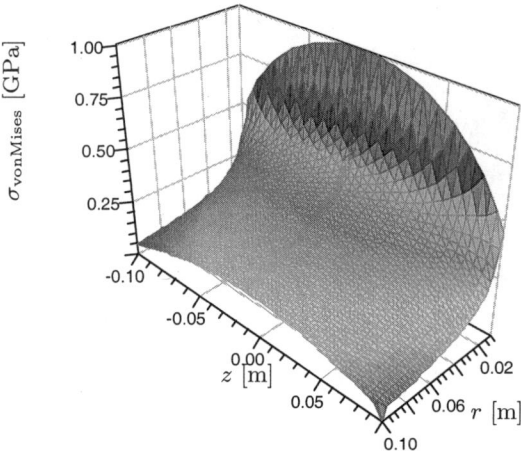

Fig. 3.11. Distribution of the von Mises stress in the cross-section of the coil from Table 3.6. Because of the rotational symmetry of the solenoid the von Mises stress has a maximum of four non-zero components and is expressed in cylindrical coordinates as

$$\sigma_{\text{vonMises}} = \sqrt{\tfrac{1}{2}\left[(\sigma_{rr}-\sigma_{\varphi\varphi})^2 + (\sigma_{rr}-\sigma_{zz})^2 + (\sigma_{\varphi\varphi}-\sigma_{zz})^2\right] + 3\sigma_{rz}^2}$$

and is symmetric with respect to the midplane. The maximum von Mises stress occurs in the midplane at the inner radius of the coil; it was adjusted to 1 GPa by selection of the proper current density in the cross-section

the axial stress σ_{zz} and finally by the shear stress σ_{rz}. This behavior gives a first hint of the quality of the estimation from Sect. 2.3: The axial stress has an effect of about half of the radial stress, and so neglecting the shear component is surely justified.

Of course we have to issue a warning here: the relative magnitudes of the different stress components may be different for very long or very flat coils. In principle the above statement regarding the relative contributions of the stress components holds only for the calculated coil; for a more general statement one would have to calculate several coils in order to have at least a parameter region of outer radii and coil heights. But even then one has to make statements like '...only holds for coils in this parameter space'. This is the main disadvantage of numerical calculations like the FEM method; they are exact in the investigated parameter space, but extrapolations and generalizations should be avoided.

The von Mises stress has a maximum value in the midplane $z = 0$ at the inner surface of the coil. For a fixed radius the highest von Mises stress occurs also in the midplane. Since the hoop stress constitutes the main component of the von Mises stress, it shows the same behavior. The radial stress σ_{rr} disappears at the inner radius and at the outer radius of the coil. Its maximum value occurs in the midplane near the inner surface of the cross-section. Close to the outer radius the radial stress becomes slightly negative in this example.

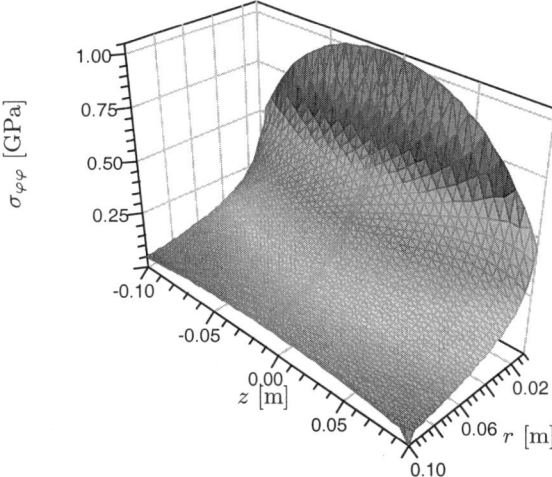

Fig. 3.12. The hoop stress $\sigma_{\varphi\varphi}$ in the cross-section of the coil. It is the main contribution to the von Mises stress. The radial stress σ_{rr}, the axial stress σ_{zz} and the shear stress σ_{rz} contribute with decreasing magnitude (see Figs. 3.13–3.15)

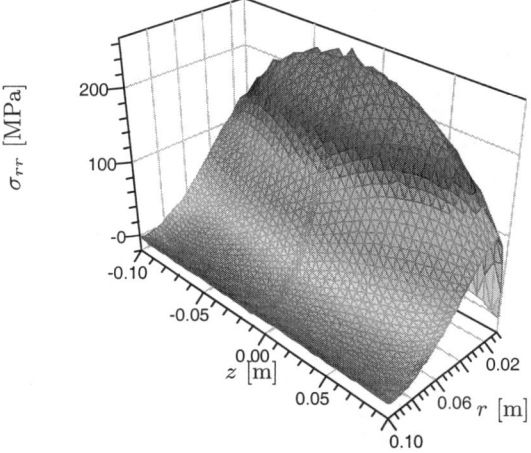

Fig. 3.13. Radial stress σ_{rr} in the coil cross-section. Note the different range of the stress coordinate! Except for a small region near the outer radius, the radial stress is always positive: two neighboring elements in the cross-section tend to detach in the radial direction. One says the radial stress in this region shows radial expansion. In the small region with negative radial stress one speaks of radial compression

This means that except for a small outer part the coil is subjected to a radial expansion: two radially consecutive areas of the cross-section try to separate radially from each other. The axial stress is by definition zero at the upper and lower surfaces of the cross-section; it is positive in an inner

112   3. Numerical Simulations

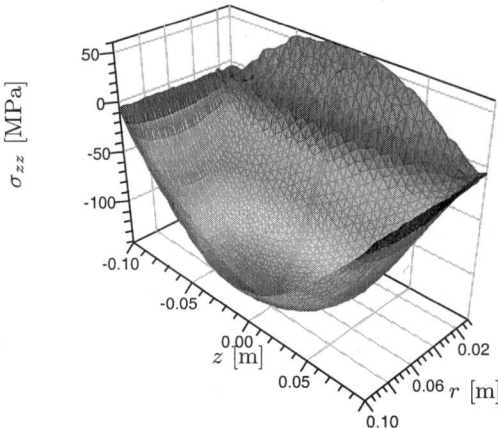

Fig. 3.14. Distribution of the axial stress σ_{zz} in the cross-section. Over most of the coil the axial stress is negative (axial compression), and only in a small region close to the inner radius is it positive (axial expansion)

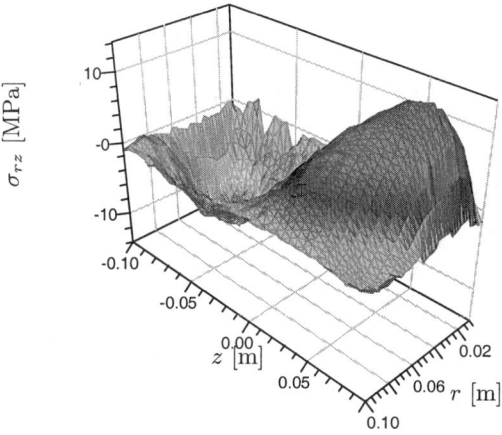

Fig. 3.15. Distribution of the shear stress σ_{rz} along the cross-section. Note the small range for the stress axis. The peaks and fluctuations are an expression of the numerical accuracy of the calculation

part of the cross-section (axial expansion) and negative in the outer part of the cross-section (axial compression). Finally, the shear stress component σ_{rz} is shown in Fig. 3.15. Basically it should be zero on all four sides of the cross-section and along the midplane. The figure shows some strange peaks and fluctuations, however. These represent the numerical accuracy of the numerical calculation.

Finally, the radial and axial dislocations a_r and a_z are given in Figs. 3.16 and 3.17. Due to the Lorentz forces the coil cross-section gets deformed; the

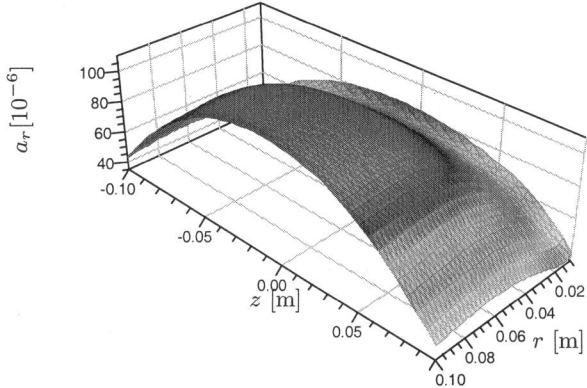

Fig. 3.16. Relative radial dislocation a_r in the coil cross-section. Due to the Lorentz forces each element in the cross-section increases its radius; the increase is highest in the midplane on the outer radius. The overall deformation causes the coil to be radially enlarged, with a 'belly' in the midplane

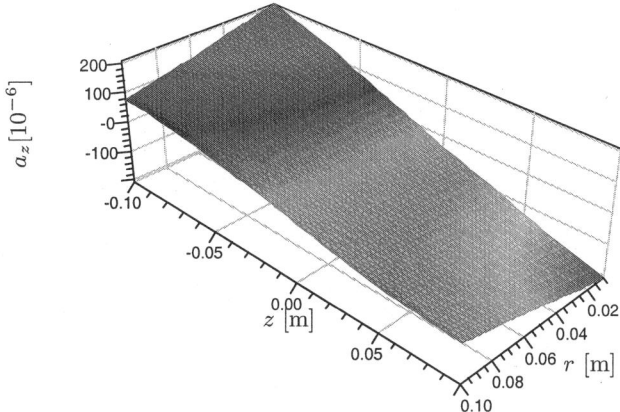

Fig. 3.17. Axial dislocation a_z in the coil under influence of the Lorentz forces. The coil gets compressed in the axial direction; the rate is highest on the inner surface of the coil at the top and the bottom

general movements are an expansion in the radial direction and a compression in the axial direction. Both movements are sketched schematically in Fig. 3.18. The deformation can be described as an overall increase in the radial direction with a 'belly' in the midplane of the coil; the axial compression causes a kind of tilting of the upper and lower side of the coil.

3.2.2 'Search' for an Optimal Coil

In this chapter we describe the 'search' for an optimal coil with the aim of finding the configuration which generates the highest field in the center under

114   3. Numerical Simulations

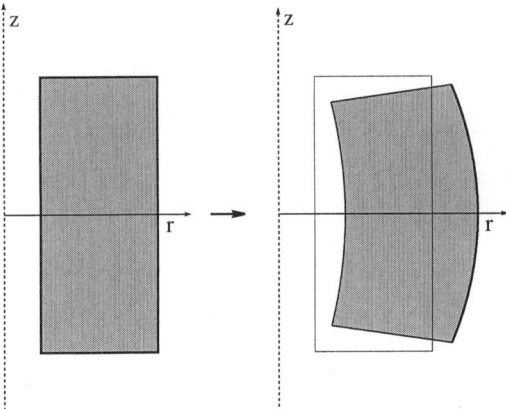

Fig. 3.18. Schematic drawing of the effect of the radial and axial dislocations from Figs. 3.16 and 3.17 on the shape of the coil. On the left hand-side the situation without current is shown; on the right side the deformation under the influence of the Lorentz forces is drawn (the effect is of course exaggerated). The rectangular cross-section gets expanded in the radial direction and compressed in the axial direction. The deformation depends on the position in the cross-section. The final shape is no longer rectangular, but has a 'belly' in the midplane and tilted upper and lower sides

Table 3.7. Fixed data for the search for an optimal wire-wound coil (Sect. 3.2.2). These data form the necessary input for the FEM code [119]. Numerous calculations were then performed for a range of outer radii and heights. The outer radius was changed from 50 mm to 300 mm in steps of 50 mm. The coil height ranged from 100 mm to 600 mm in steps of 100 mm

inner radius	10	mm
Young's modulus	135	GPa
Poisson ratio	0.34	
maximal von Mises stress	1	GPa
current density	const.	adapted to reach 1 GPa

the constraint of a maximum von Mises stress of 1 GPa. For this purpose numerous calculations like the one in the previous section were performed. Besides the fixed boundary condition a few other coil data were kept constant (see Table 3.7). These were a fixed inner radius of 10 mm and a material with a Young's modulus of 135 GPa and a Poisson ratio of 0.35. The current density in the rectangular cross-section was assumed to be constant, and its value was adjusted in order to reach a maximum von Mises stress in the coil of 1 GPa. For the search the varied parameters were the outer radius a_2 and the height h of the coil. The outer radius was increased from 50 mm to 300 mm

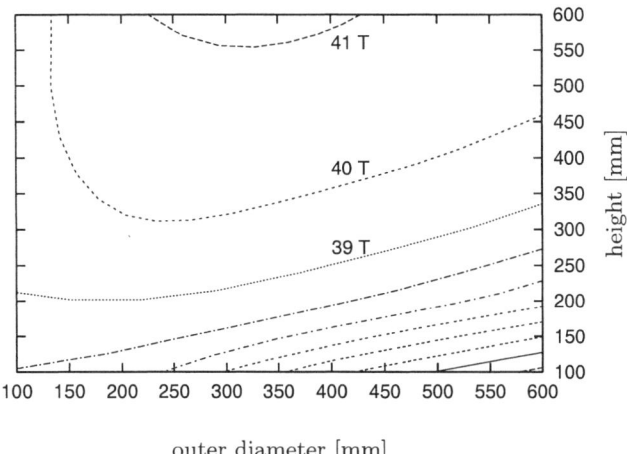

Fig. 3.19. Central magnetic field for wire-wound coils with rectangular cross-section. All coils have an inner radius of 10 mm and a constant current density, which was adjusted to get a maximum von Mises stress in the coil of 1 GPa. The dependence of the central field as a function of the outer diameter and the coil height is shown. The distance between contour lines is 1 T (for reasons of clarity only a few are labelled). The calculations were performed with the method of finite elements, the material in the cross-section was assumed to be homogeneous, isotropic and linear, the Young's modulus was 135 GPa and the Poisson ratio 0.34. In the investigated region the maximum field of 41.2 T is produced by a coil with an outer diameter of about 300 mm and a height of 600 mm

in steps of 50 mm; the height ranged from 100 mm to 600 mm in steps of 100 mm. This grid defined a total of 121 different coil configurations.

The central fields of these calculations are shown in Fig. 3.19 as a contour plot. The general tendency in the parameter space shown is an increase of the central field with increasing coil height. As indicated by the increasing gap between following contour lines (e.g. 39 – 40 – 41 T) there is some kind of saturation visible. Indeed, the estimation with the (infinite) long coil from Sect. 2.3 shows that the maximum achievable field under the condition of a maximum von Mises stress of 1 GPa is 41.5 T. Obviously the finite element calculations converge towards that value, when the height gets bigger and bigger.

On the other hand, a statement like 'the central field increases with increasing outer radius' is wrong, as the contour line to 40 T clearly shows. However, if we would have shown only the data to heights up to 200 mm, one might have come to that conclusion. So we want again to issue the warning that generalizations from numerical calculations into unknown parameter regions should be made with great care.

In the investigated parameter space the coil with the highest field in the center would have an outer diameter of about 300 mm and a height of 600 mm,

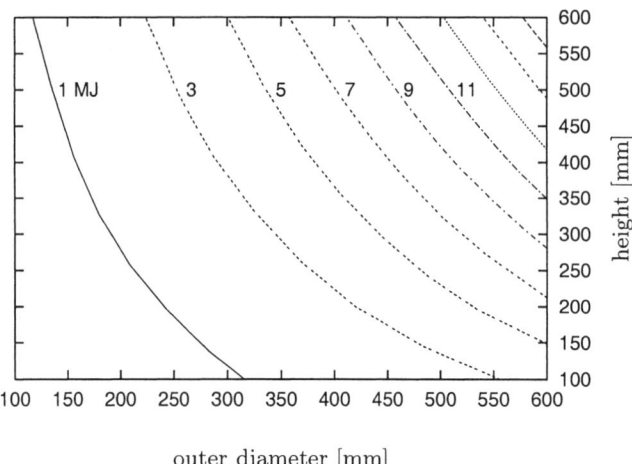

Fig. 3.20. Magnetic energies associated with the coils from Fig. 3.19. In the calculated region the magnetic energy increases with increasing outer diameter and increasing height of the coil. The coil generating the maximum field needs about 5 MJ (outer diameter 300 mm, height 600 mm). Increasing the outer diameter of this 'optimal coil' makes a coil which generates a lower field, but which needs more magnetic energy to energize the coil to its maximum von Mises stress of 1 GPa. The advantage of such a coil would be a decrease in the current density, as can be seen in Fig. 3.21

with a central field of 41.2 T. The necessary current density is calculated as $2.444 \times 10^8 \, \mathrm{A\,m^{-2}}$, and the associated magnetic energy as 5.0 MJ. The contour plot of the magnetic energies is shown in Fig. 3.20. One can save a considerable amount of magnetic energy for a modest decrease of the central field by decreasing the height of the coil. For instance, a coil with a height equal to the outer diameter of 300 mm would still produce 39.9 T, but needs only 2.4 MJ, which is about one half of that needed for the 'optimal coil'.

Finally, the contour plot of the current density of the various coils is shown in Fig. 3.21. In the investigated region the current density decreases with increasing outer radius and increasing coil height. If we assume the same conductor material for all the coils, then we see from $j^2 \, t_{\mathrm{pulse}} = \mathrm{const}$ that the bigger coils (outer diameter and height) have longer pulse times. From Fig. 3.20 we also see that such a coil needs a lot of magnetic energy.

In the next section we take a subset of the calculations made here, namely those where the outer diameter and height of the coil are equal, and compare them with the analytical calculations from Sect. 2.3.

3.2.3 Comparison with Analytical Methods

In this chapter we compare some of the finite element calculations of the previous section, namely those where the outer diameter and height of the

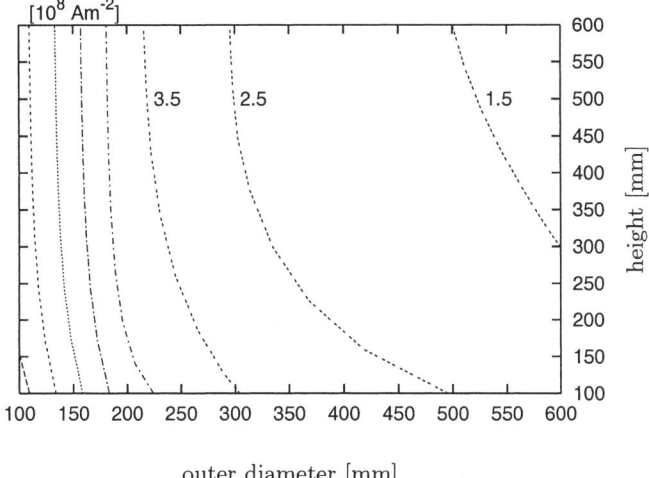

Fig. 3.21. Current densities for the finite element calculations of Figs. 3.19 and 3.20, which are necessary for achieving a maximum von Mises stress in the coil of 1 GPa. The current density decreases with increasing outer diameter and increasing height of the coil. This means that big coils allow for the longest pulse times, since $j^2\, t_{\text{pulse}} = \text{const}$; the same conductor material is assumed here. Again for reasons of clarity, only a few contour lines have labels; the distance between two contour lines is $10^8\,\text{A}\,\text{m}^{-2}$

Table 3.8. Fixed data for the comparison of a wire-wound coil with the method of finite elements and an analytical estimation. The outer diameter and height of the coil are assumed to be equal; the outer radius was changed from 50 mm to 300 mm in steps of 25 mm

inner radius	10	mm
Young's modulus	135	GPa
Poisson ratio	0.34	
maximal von Mises stress	1.0	GPa
current density	constant	adapted to reach 1.0 GPa

coil are equal, with the estimations from Sect. 2.3, where we used the trick of extending the coil to infinity in the axial direction in order to calculate the fields and stresses in an analytical way.

The constant data for each coil are summarized in Table 3.8, the variable parameter is now the outer diameter of the coil only. In Fig. 3.22 the central fields for the finite element calculations and the analytical estimation are shown as a function of the outer radius of the coil. Both graphs show the same behavior: with increasing outer radius they saturate to a certain value, which is determined by the allowed maximum von Mises stress, i.e. the yield strength of the coil structure. In the range of outer diameters shown the

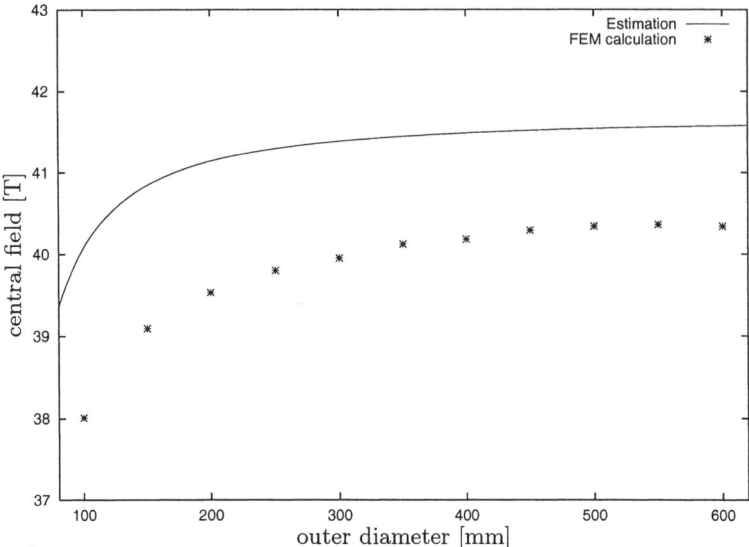

Fig. 3.22. Central fields calculated by the finite element method and the analytical estimation from Sect. 2.3 as a function of the outer diameter of the coil. The height of the coil is always taken as equal to the outer diameter; the fixed coil data can be found in Table 3.8. Both graphs show a saturation of the central field with increasing outer diameter. The saturation value depends on the admitted yield strength of the coil structure. The analytical method overestimates the field by about 5%

analytical method with the infinitly long coil results in an overestimation of the central field of about 5% compared to the calculations with the finite element method.

Whereas with increasing outer diameter the increase in the central field gets smaller and smaller, there occurs a drastic increase in the necessary magnetic energy, which can be seen in Fig. 3.23. Over the parameter range for the outer radius shown the estimation from Sect. 2.3 gives about a 15% higher value for the magnetic energy.

Finally, Fig. 3.24 shows the current densities for the two calculation methods. Here the values for the finite element calculations are greater than those from the estimation. The difference is around 10%. In both cases the current density decreases with increasing outer diameter.

The comparison of both calculation methods shows that the estimation from Sect. 2.3 with the infinitly long coil trick produces similar results for the central field, the necessary magnetic energy and the current density. The central field is overestimated by about 5%, the magnetic energy by about 15% and the current density is underestimated by roughly 10%.

On the other hand, the total computing time for the finite element calculations for Figs. 3.22–3.24 lies in the range of one hour, whereas the analytical estimation is done almost instantly. A common sense approach for finding a

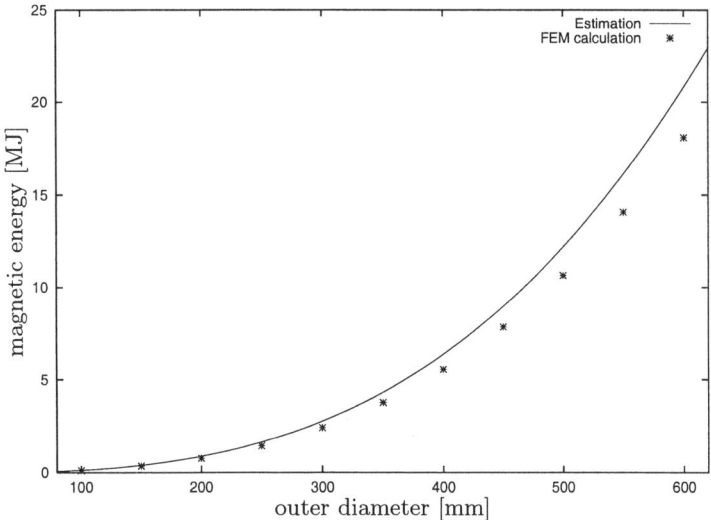

Fig. 3.23. Necessary magnetic energy as a function of the outer diameter of the coil. The magnetic energy increases with increasing diameter of the coil; the estimation gives about 15% higher values than the finite element calculation for the same outer diameter

certain coil design would therefore start with an analytical estimation. Once a 'good' point in the parameter space is found, one will perform numerical calculations around this 'optimal' solution. We demonstrate this technique in the next section.

3.2.4 Two-Coil System

In this section we describe the design process for an arrangement of two coils. The situation is sketched in Fig. 3.25. Two coils with rectangular cross-section are nested into each other, separated by a gap g. Each coil is treated as a massive coil, i.e. the mechanical forces are transmitted in the radial direction. The current densities in both coils are assumed as constant; however, each coil can have different values for the current density. The inner and outer coil allow a maximum von Mises stress of σ_1 and σ_2, respectively.

In the example given here we vary the maximum von Mises stress σ_1 of the inner coil and adjust the separation radius a_m to an optimal value, so that a maximum central field will be generated. The fixed parameters are the inner radius a_{11} of the inner coil, the outer radius a_{22} of the outer coil and the two coil heights h_1 and h_2. The maximum von Mises stress in the outer coil is fixed as $\sigma_2 = 1.0\,\text{GPa}$; for the maximum von Mises stress of the inner coil we investigate the values $\sigma_1 = 1.0, 1.5$ and $2.0\,\text{GPa}$. The coil data are summarized in Table 3.9.

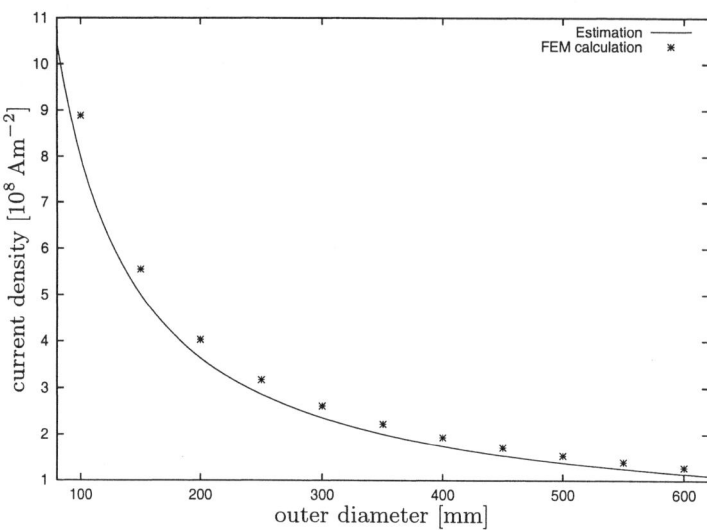

Fig. 3.24. Comparison of the current density for the finite element calculation and the estimation with the infinitely long coil from Sect. 2.3. The current density was adjusted to give a maximum von Mises stress in the coil of 1.0 GPa. In both cases the current density shows the same behavior: a decreasing value for the current density with increasing outer diameter. Because $j^2 \, t_{\text{pulse}} = \text{const}$, this means that bigger coils have longer pulse lengths. In the diameter range shown the estimated current density lies about 10% below the values from the finite element calculations

We start first with the estimation from Sect. 2.3.1, which treats an infinitely long two-coil system, i.e. the two heights h_1 and h_2 of the real system are neglected all together. The analytical calculation gives for the three values of $\sigma_1 = 1.0$, 1.5 and 2.0 GPa a separation radius of $a_{\text{m}} = 43.4$, 49.0 and 53.8 mm; further results for the central fields and the magnetic energies are given in Table 3.10.

The three values for the separation radius a_{m} define the starting point for more accurate calculations with the method of finite elements. The gap between the inner and outer coil is assumed to be $g = 5$ mm; the material properties of both coils are furthermore taken as homogeneous and isotropic. For Young's modulus we use $E = 135$ GPa and for the Poisson ratio $\nu = 0.34$. The current density in both coils is manually adjusted so that the desired values for the maximum von Mises stress in each coil are reached. Several more calculations are performed around the estimated separation radii a_{m} from Table 3.10. The results of the finite element calculations are given in Table 3.11.

Again, the analytical estimation and the finite element calculations show the same behavior: the central field increases with increasing value of the allowed stress level σ_1 in the inner coil. The analytical method gives about 5% higher values for the central fields and the separation radius is estimated

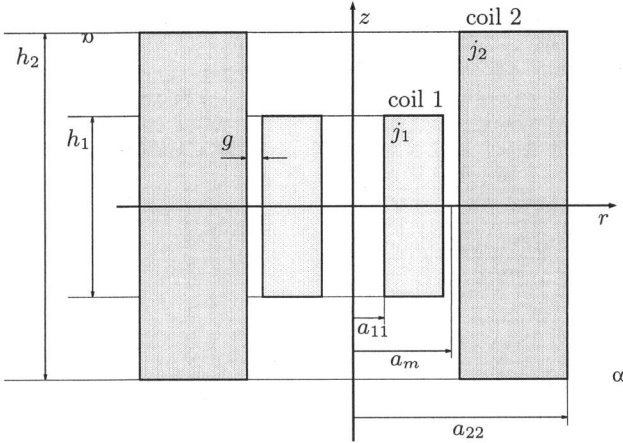

Fig. 3.25. Setup for a two-coil system, consisting of two nested coils with rectangular cross-section and with constant current densities j_1 and j_2. The two heights h_1 and h_2 of the coils and the innermost and outermost radii a_{11} and a_{22} are assumed to be constant, and the parameter to be optimized is the separation radius $a_{\rm m}$. The aim is to find the optimal separation radius, for which the two-coil system generates the highest possible field under the boundary condition that the von Mises stress in both coils stays below the two given limits σ_1 and σ_2, respectively. The gap g between the two coils is also taken as constant; numerical values for a simulation are given in Table 3.9

Table 3.9. Fixed data for the two-coil system. The only free parameter is the separation radius $a_{\rm m}$. At the optimal separation radius the coil system will generate the highest field complying with the boundary condition of finite stresses in both coils

inner radius, inner coil	a_{11}	10	mm
outer radius, outer coil	a_{22}	200	mm
height of inner coil	h_1	200	mm
height of outer coil	h_2	400	mm
gap between coils	g	5	mm
max. stress in inner coil	σ_1	1.0, 1.5 and 2.0	GPa
max. stress in outer coil	σ_2	1.0	GPa
Young's modulus	E	135	GPa
Poisson ratio	ν	0.34	
current densities	constant in each coil		

as being somewhat too large. The finite element calculations show the maximum central field occurring always towards lower values of the respective separation radius $a_{\rm m}$. The differences in the generated fields are rather small, however, an important feature when it comes to the manufacture of the coil

Table 3.10. Estimation of the two-coil system from Table 3.9 performed with the analytical method from Sect. 2.3.1. The calculated separation radii define the starting points for finite element calculations. The generated field depends strongly on the the allowed stress level σ_1: the higher this value is, the more field can be generated. Note the dramatic increase for the case of $\sigma_1 = 1.0\,\text{GPa}$ when compared to a single coil with the same inner and outer radii and the same height as the two-coil system; contrast 55.44 T in table vs. 41.49 T in Fig. 3.22

σ_1 [GPa]	a_m [mm]	B_0 [T]	W_m [MJ]
1.0	43.4	55.44	7.87
1.5	49.0	61.91	8.25
2.0	53.8	67.86	8.63

Table 3.11. Results of the analytical estimation and finite element calculations for the two-coil system of Table 3.9. The estimated values for the central field B_0 are about 5% too high; also the estimated separation radius a_2 tends to be somewhat too large. The estimated values for the magnetic energy W_m also are a few percent higher than those calculated by the FEM method. For completeness the current densities j_1 and j_2 in the inner and the outer coil are also shown

σ_1 [GPa]	a_m [mm]	B_0 [T]	W_m [MJ]	j_1 [$10^8\,\text{A}\,\text{m}^{-2}$]	j_2 [$10^8\,\text{A}\,\text{m}^{-2}$]	Method
1.0	43.4	55.44	7.87			estimation
	37.0	53.40	7.28	5.13	2.21	FEM
	40.0	53.47	7.47	4.69	2.25	FEM
	43.0	53.45	7.67	4.32	2.28	FEM
	46.0	53.35	7.86	4.00	2.31	FEM
1.5	49.0	61.91	8.25			estimation
	43.0	59.83	7.79	6.03	2.28	FEM
	46.0	59.85	8.00	5.59	2.32	FEM
	49.0	59.79	8.21	5.21	2.35	FEM
2.0	53.8	67.86	8.63			estimation
	45.0	65.71	8.06	7.21	2.31	FEM
	48.0	65.75	8.28	6.71	2.34	FEM
	51.0	65.72	8.51	6.28	2.38	FEM
	54.0	65.49	8.73	5.87	2.41	FEM

system, because this means the system is not very sensitive with respect to manufacturing tolerances.

The case of $\sigma_1 = 1.0\,\text{GPa}$ deserves special attention, when one compares the two-coil system with a monolithic coil of the same geometric dimensions. The finite element calculation of the single coil (see Fig. 3.22) allows us to generate a central field of 40.18 T, whereas the two-coil system can produce about 53.47 T. This demonstrates clearly the superiority of the two-coil system over the one-coil system. The only drawback is a decrease of the pulse length, since the current density in the inner coil of the two-coil system is higher than that of the single-coil system (4.69 vs. $1.93 \times 10^8\,\mathrm{A\,m^{-2}}$).

A further optimization of the current density and the overall cross-section of the coil can be achieved by dividing the coil not in two but three, four ...parts, ultimately leading to a polyhelix coil. This is accompanied by a more complicated mechanical and electrical design, since the single coils have to be fixed and aligned to each other and electrical joints have to be made between them. The two-coil system is the first step towards a polyhelix coil with the advantage of a relatively simple mechanical and electrical design.

3.3 Plastic Deformation

In this chapter we present numerical simulations of wire-wound coils where we also incorporate plastic deformations. As outlined in Sect. 1.2.6 the plastic behavior of a material can be described by a nonlinear relationship between the von Mises strain and the von Mises stress, both of which are also referred to as the effective stress and strain.

For numerical purposes – the main reason is to achieve an acceptable runtime – this relationship is approximated with a piecewise linearization. The numerical algorithm then repeatedly calculates the problem with a stepwise increase of the load up to the final load. During each step the effective strain of each volume element is calculated, and depending on the stress-strain relation, a corresponding von Mises stress and Young's modulus is associated with the volume element. A very fine linearization therefore requires very small incremental load steps, since 'jumps' in the $\sigma(\varepsilon)$ curve should not occur. For reasons of accuracy every linearized region of the $\sigma(\varepsilon)$ curve should be reached in ascending order.

In the case of a coil the load on the coil consists of the Lorentz forces, i.e. the algorithm simulates the loading of the coil by increasing the current through the coil in small steps from zero until the desired current value is reached.

For the calculation presented here we take a rather simple approximation for the stress-strain relation, decomposing it into only two linear regions, the first one representing the linear elastic regime followed by a plastic regime (see Fig. 3.26).

The coil type we investigate here is a wire-wound coil with internal reinforcement. For the wire we use hard copper and for the reinforcement either Zylon (Sect. 3.3.1) or S2-glass (Sect. 3.3.2). The material data are taken from [81, 120–125] and are summarized in Table 3.12. We allow plastic deformation in the copper, whereas the reinforcement always stays in the elastic regime.

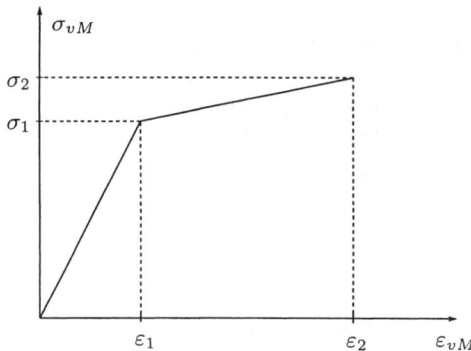

Fig. 3.26. The linearization scheme used for describing the behavior of hard copper. For an effective strain below the elastic limit $\varepsilon_1 = 0.16\%$, the copper is in its elastic regime. The plastic regime follows and at an effective strain of $\varepsilon_2 = 2.16\%$ we assume the material to fail ($\sigma_1 = 200\,\text{MPa}$, $\sigma_2 = 400\,\text{MPa}$)

Table 3.12. Data for conductor and reinforcement for numerical calculations of coils allowing plastic deformation in the conductor. Two simulations were performed: one uses Zylon, the other S2-glass as reinforcement

Material			
Copper	$\varepsilon_1 = 0.2\%$	$\sigma_1 = 250\,\text{MPa}$	elastic limit
	$\varepsilon_2 = 2.5\%$	$\sigma_2 = 400\,\text{MPa}$	break
	$E_0 = 125\,\text{GPa}$	$E_1 = 6.5\,\text{GPa}$	elastic modulus
Zylon	$\varepsilon_{\max} = 2.5\%$	$\sigma_{\max} = 5.8\,\text{GPa}$	data for fiber-epoxy
	$E_0 = 230\,\text{GPa}$		epoxy material
S2-glass	$\varepsilon_{\max} = 3.1\%$	$\sigma_{\max} = 2.5$	data for fiber-epoxy
	$E_0 = 80\,\text{GPa}$		epoxy material

In both cases the the coil had the same geometric dimensions, summarized in Table 3.13. The inner four conductor-reinforcement layers are followed by a 0.2 mm thick layer of a very soft material like Teflon tape. This allows the layers to separate from each other. Calculations performed without that Teflon tape show the radial stress component σ_{rr} to be positive in the inner region of the coil. A positive radial stress component means that the outer layers pull on the inner layers, thereby causing additional stress in the inner layers. By putting Teflon gaps as layer separations in the regions with $\sigma_{rr} > 0$ we therefore allow a separation of the layers, and no outer layers can pull on the inner ones anymore. This causes a significant decrease in the von Mises stress in the inner layers. Basically the coil then behaves like a polyhelix coil in its inner region!

Table 3.13. Data for the coil for the numerical calculations with plastic deformation in the copper conductor

inner radius	5 mm	
outer radius	44 mm	Zylon
	44 mm	S2-glass
height	100 mm	
thickness of conductor layers	2 mm	
number of layers	8	
current density radius	4.4×10^9 A m^{-2}	Zylon
	3.3×10^9 A m^{-2}	S2-glass

3.3.1 Zylon Reinforcement

Zylon is the trade name of a high-strength fiber (see [125]). This fiber has an elastic modulus of up to 280 GPA and an ultimate tensile strength of 5.8 GPa. A further advantage is the very high packing factor of the fiber in a fiber-epoxy laminate, caused by the deformation of the fiber cross-section when wound under high tension. Microscopic inspections of such laminates show an almost hexagonal cross-section of the fiber [126].

For the purpose of the numerical calculations we assume for the Zylon-epoxy reinforcement an ultimate tensile strength of 3 GPa and an elastic modulus of 230 GPa [127]. Taking the maximum values for the stresses and the strains in the copper and the reinforcement into account then the coil can sustain a maximum current density of 4.4×10^9 A m^{-2} and a central field of about 80 T is generated. The magnetic energy in the coil at this current reaches 264 KJ. Figure 3.27 shows the von Mises stress in the midplane of the coil. The stresses in the copper conductor are limited to about 400 MPA due to the plastic deformation, and the stresses in the Zylon reinforcement are lower than the desired 3 GPa. Except for the second layer the stresses in the Zylon are somewhat lower; by decreasing the thickness of the respective layer one could put them also up to the limit of 3 GPa. On the other hand, the coil as represented here works on the safe side.

The equivalent strain ε_{vM} along the midplane of the coil is shown in Fig. 3.28. It stays everywhere within the allowed limits; for the copper conductor, in particular, it never exceeds the allowed limit of 2.5%. The first and second copper layer reach this limit, however, which means that the coil works at its limits.

The axial and radial displacements a_z and a_r, respectively, are shown in Figs. 3.29 and 3.30. The maximum values are 0.27 mm for the radial and 0.25 mm for the axial displacement. Also nicely seen are the effects of the Zylon tape gaps in the coil, this causes the four steps in the graphs of the radial and axial displacement.

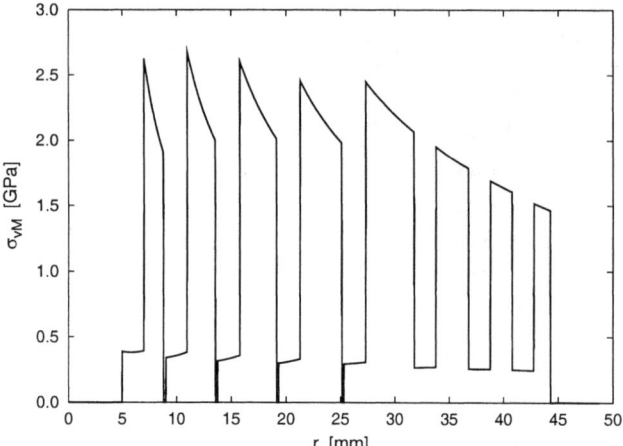

Fig. 3.27. von Mises stress as function of the radius in the midplane of a coil with Zylon reinforcement. In all Zylon layers the stress obeys the requirement of staying below 3 GPa; in the copper layers a maximum stress of 0.4 GPa, the assumed yield stress is not exceeded. The Teflon layers after the four innermost layers show up as drops in the stress

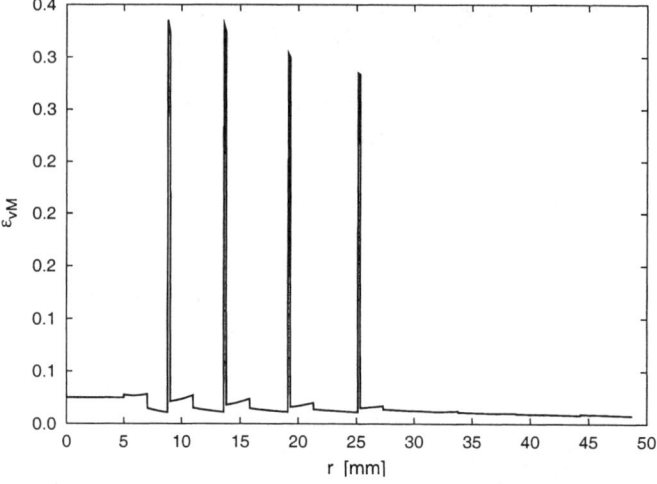

Fig. 3.28. von Mises strain as function of the radius in the midplane of a coil with Zylon reinforcement. In the first and second copper layers the strain almost reaches the limit of 2.5%, but all other layers are uncritical

The coil from reference [127] was designed for 78 T at a peak stress level of $\sigma_{vM} = 3.2$ GPa using proprietary software. A coil wound corresponding to this design has survived 74.8 T without coil failure so far. The slight difference between our calculation and those from [127] might be due to the chosen

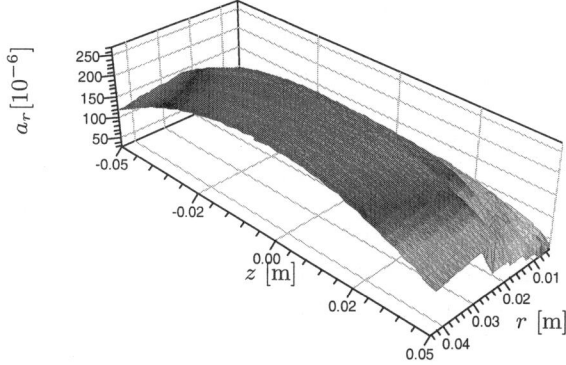

Fig. 3.29. Distribution of the radial displacement a_r in the coil cross-section. Again one can see the steps in the displacement due to the Teflon gaps

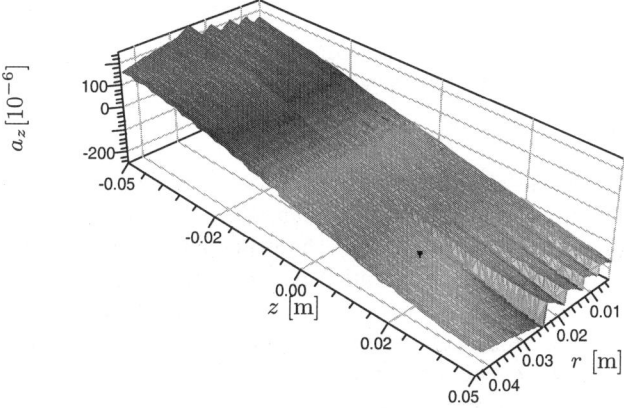

Fig. 3.30. Distribution of the axial displacement a_z in the coil cross-section. Nicely seen are the jumps in the displacement due to the Teflon tape. At full field the coil contracts in the axial direction by about 0.4 mm

calculation methods as well as slightly different material properties. Finally, the real coil might not be what one calculates and, due to Murphy's law, will never be as good as any calculation.

3.3.2 S2-Glass Reinforcement

Now the Zylon is substituted with S2-glass. Because of the much lower elastic modulus of the S2-glass-epoxy matrix the copper conductor reaches its maximum von Mises stress and strain at a much lower current density of $3.3 \times 10^9 \, \mathrm{A\,m^{-2}}$, which means the coil can now produce a central field of only about 60 T with a magnetic energy of 149 KJ. The von Mises stress in the coil midplane is shown in Fig. 3.31. The maximum value of 400 MPa for the

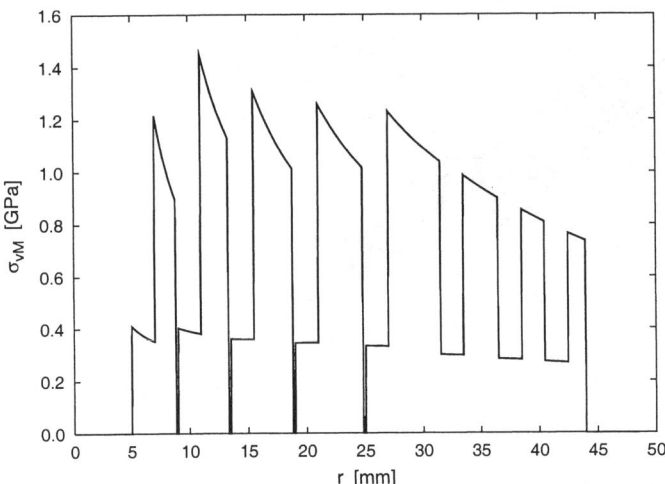

Fig. 3.31. von Mises stress as a function of the radius in the midplane of a coil with S2-glass reinforcement. The two innermost copper layers reach the stress limit of 0.4 GPa, whereas all S2-glass reinforcement layers see only uncritical stresses

copper is reached in the innermost conductor layer, and in the reinforcement there occurs a maximal stress of 145 MPa. This means that the stress capability of the S2-glass cannot be used at all. On the other hand, the strain in the conductor is already at its limit of about 2.5% (see Fig. 3.32), and so making the S2-glass layer thinner in order to increase the stress in the reinforcement layer and make use of the full strength capability of the material is no option since the conductor will then fail. Figures 3.33 and 3.34 finally show the radial and axial displacement of the coil; they are similar to the figures for the Zylon coil.

3.3.3 Conclusion

This chapter has demonstrated the use of a nonlinear relationship between the equivalent stress and strain in order to describe the elastic and plastic behavior of a material. For numerical purposes this relationship is often approximated by a piecewise linearization.

The study showed the benefit of introducing Teflon tape between the innermost layers so that the layers can be separated from each other; the inner part of the coil is then equivalent to a polyhelix coil.

The maximum field that can be reached depends on the maximum allowable values for either stress or strain in the coil. The two different reinforcement materials show that we must also look at the elastic modulus of the conductor and the reinforcement. Ideally the elastic modulus of the fiber-epoxy reinforcement should be grater than the elastic modulus of the conductor material. Otherwise, the reinforcement is not used at its full potential at

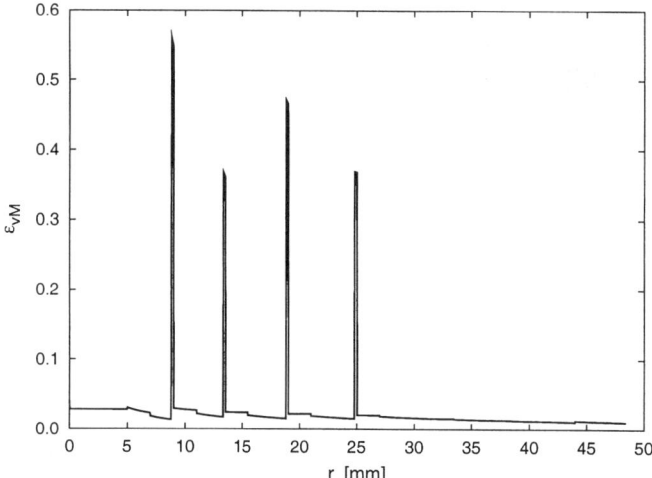

Fig. 3.32. von Mises strain as a function of the radius in the midplane of a coil with S2-glass reinforcement. In the first and second copper layers the strain exceeds the strain limit of 2.5%; all other layers are uncritical

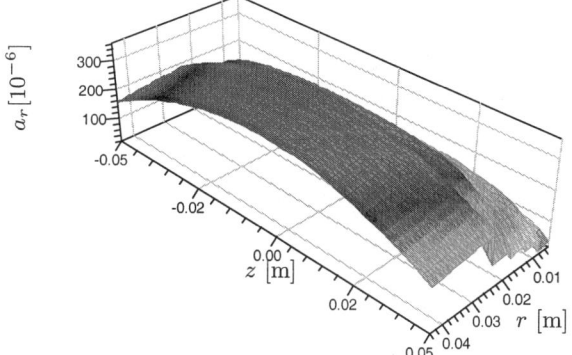

Fig. 3.33. Distribution of the radial displacement a_r in the coil cross-section. Again, the Teflon gaps cause jumps

all. These criteria are fulfilled for a Zylon-epoxy reinforcement. Another good material would be graphite fibers; unfortunately these are slightly conductive so that insulation of the conductor can become an issue.

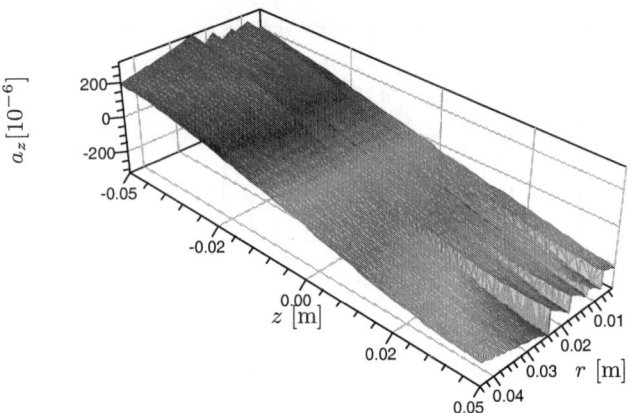

Fig. 3.34. Distribution of the axial displacement a_z in the coil cross-section

4. Pulsed Field Facilities

The energy source for a pulsed coil system has to comply with certain requirements regarding stored energy and peak power delivered to the coil. A quick estimation can be performed with the methods derived in Sect. 2.2. A constant current density coil with inner bore diameter of 10 mm and coil parameters $\alpha = \beta = 20$ made from a hypothetical material with a tensile strength of 2.2 GPa could produce a central field of 100 T, requiring a magnetic energy of 4.6 MJ. If we assume a rise time of the pulsed field of 5 ms, then the pulsed power source would need to deliver an average power of about 1 GW.

The storage of energy can be realized in the form of kinetic energy in flywheel generators, as chemical energy in batteries, as magnetic energy in an inductor or as electrical energy in a capacitor. In this chapter we derive the basic equations for pulsed coil facilities realized with these storage techniques. Regarding the type of the equation there emerge only two types: one describes the capacitive energy storage and the other one the inductive energy storage. A battery system, for instance, can be simulated as a very large capacitor with a value of a few farads.

We concentrate first on the capacitive energy storage method and begin with a circuit with constant values for the circuit elements. The high-field coil, in particular, is treated as having a constant resistance. In the next step towards the simulation of a real coil we incorporate the resistance increase of the high-field coil due to ohmic heating. Further refinements follow with addition of the magnetoresistance. Finally, we describe the influence of eddy currents.

The second part of this chapter deals with the inductive energy transfer. The analysis is restricted to constant circuit parameters. It starts with ideal coils, i.e. coils having no resistance, and then moves on to more realistic coils with internal resistance. Finally, some possibilities for increasing the efficiency of the energy transfer are discussed.

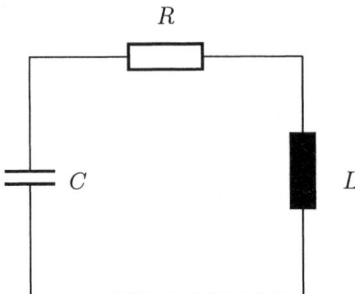

Fig. 4.1. Principal circuit of a capacitor C discharged into a coil, consisting of an inductance L and a resistance R

4.1 Capacitive Energy Storage

4.1.1 Ideal RLC Circuit

The main idea here is to discharge a capacitor into a coil. The principal circuit is shown in Fig. 4.1; it consists of the capacitor C and the coil, which is formed by an inductance L and a resistance R. At the beginning of the discharging process the capacitor is loaded with a voltage $U_{\max}$, and the current I through the magnet is zero. The differential equation which governs the behavior of the circuit is

$$\frac{1}{C}I + R\dot{I} + L\ddot{I} = 0 \,. \tag{4.1}$$

It has to be solved for the initial conditions $I(0) = 0$ and $Q(0) = CU_{\max}$. Equation (4.1) describes a damped oscillation; depending on a damping parameter d one can distinguish undercritical ($d < 1$), critical ($d = 1$) and overcritical ($d > 1$) damping. The damping parameter is defined by

$$d = \frac{R^2 C}{4L} \,. \tag{4.2}$$

With the definition of the following constants

$$\omega_0 = \sqrt{\frac{1}{LC}}, \quad \gamma = \frac{R}{2L} = \omega_0 \sqrt{d}, \quad \omega = \omega_0 \sqrt{|1-d|} \tag{4.3}$$

and

$$W_0 = \frac{1}{2} C U_{\max}^2, \quad I_0 = \sqrt{\frac{2W_0}{L}} \tag{4.4}$$

the solution for the current is

$$I(t) = I_0 \, e^{-\gamma t} \begin{cases} \frac{1}{\sqrt{1-d}} \sin \omega t & \text{for} \quad d < 1 \\ \omega_0 t & \text{for} \quad d = 1 \\ \frac{1}{\sqrt{d-1}} \sinh \omega t & \text{for} \quad d > 1 \end{cases}. \quad (4.5)$$

The time when the current reaches its first maximum and the value of this current can be written as

$$t_{\max} = \frac{1}{\omega_0} \frac{-\ln k(d)}{\sqrt{d}}, \quad (4.6)$$

$$I_{\max} = I_0 \, k(d) \quad (4.7)$$

with an attenuation function $k(d)$ defined as

$$k(d) = \begin{cases} \exp\left[-\sqrt{\frac{d}{1-d}} \arctan \sqrt{\frac{1-d}{d}}\right] & \text{for} \quad d < 1 \\ \exp[-1] & \text{for} \quad d = 1 \\ \exp\left[-\sqrt{\frac{d}{d-1}} \tanh^{-1} \sqrt{\frac{d-1}{d}}\right] & \text{for} \quad d > 1 \end{cases}. \quad (4.8)$$

For a comparison of the efficiency of the energy transfer one has to look at the magnetic energy stored in the coil. Its maximum value is calculated to be

$$W_{m,\max} = W_0 \, k(d)^2. \quad (4.9)$$

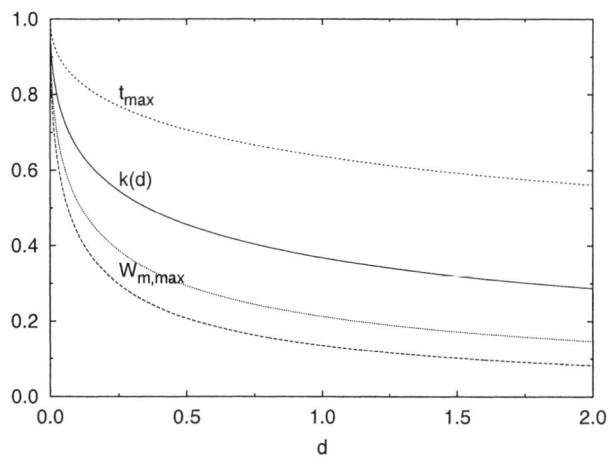

Fig. 4.2. Time and value of the maximum current and the maximum magnetic energy as a function of the damping parameter d. All three functions are normalized with their values for $d = 0$: $t_{\max}(0) = \frac{1}{\omega_0} \frac{\pi}{2}$, $I_{\max}(0) = I_0$ and $W_{m,\max}(0) = W_0$. The aperiodic damping $d = 1$ separates the region of undercritical damping $0 \leq d < 1$ and the region of overcritical damping $1 < d$. For efficient energy transfer the magnetic energy should be as high as possible, i.e. the damping d should be small

Graphs of the attenuation function $k(d)$, the time of the current maximum t_{max} and the magnetic energy $W_{m,max}$ as functions of the damping parameter d are shown in Fig. 4.2. The time and the value of the maximum current and the maximum magnetic energy have their highest overall values in the case of no damping ($d = 0$); with increasing damping parameter d they all fall strongly monotonically. For a high efficiency of the energy transfer from the capacitor into the coil the circuit should therefore be as less damped as possible.

4.1.2 Resistance Increase Due to Heating

The formalism developed in the last section was very straightforward, because all the circuit elements were regarded as being constant. In a real coil system this is no longer the case, especially because the high-field coil experiences a considerable temperature increase during the pulse. Depending on the circuit the coil can get heated up from 77 K to room temperature and above. The upper limit is set by the construction, for instance by a degradation of the wire insulation. Another concern can be large temperature gradients, since during the current pulse only the wire gets heated up and there is no considerable heat transfer into the coil's supporting structure, as for instance a glass fiber reinforcement. Temperature gradients of course are a specific problem related to the coil construction; the problems are different for a wire-wound coil than for a polyhelix coil.

As a first step towards reality we incorporate the temperature increase of the coil resistance during the discharge process of the capacitor. The governing equation is still (4.1), but it holds only for a small time interval, during which the resistance is regarded as constant. Since the change of the resistance with time and temperature is generally a complicated function, we have again to resort to numerical methods, one of the standard ones being the Runge-Kutta methods [128]. For the simulation performed here we selected a Runge-Kutta method of 4th order. The pseudo-algorithm is as follows. The simulation starts with the initial values of temperature T_0, current I_0 and voltage U_0 of the capacitor:

$T_0 = 77\,\text{K}$,
$I_0 = 0\,\text{A}$,
$U_0 = U_{max}$.

With a suitable time interval Δt, which for instance can be chosen as 1/100 of the period of the undamped LC-circuit ($\Delta t = (2\pi\sqrt{LC})/100$), the following iteration process has to be performed:

1. Calculate the temperature increase due to ohmic heating during the time interval Δt. Denoting the specific heat capacity taken at the temperature T_n by c_n and the mass by m, we obtain

$$T_{n+1} = T_n + \frac{R_n I_n^2}{c_n(T_n)\, m} \Delta t \,.$$

2. Calculate the resistance at the new temperature:

$$R_{n+1} = R(T_{n+1}) \,.$$

3. Now apply the Runge-Kutta method, using four intermediate constants $k_1 \ldots k_4$ and the derivative $\dot{I}$ of the current:

$$k_1 = \Delta t\, \dot{I}(I_n)$$
$$k_2 = \Delta t\, \dot{I}(I_n + \frac{k_1}{2})$$
$$k_3 = \Delta t\, \dot{I}(I_n + \frac{k_2}{2})$$
$$k_4 = \Delta t\, \dot{I}(I_n + k_2)$$
$$I_{n+1} = I_n + \frac{k_1 + 2k_2 + 2k_3 + k_4}{6}$$
$$U_{n+1} = U_n - \frac{I_{n+1}}{C} \Delta t \,.$$

The derivative of the current is defined as

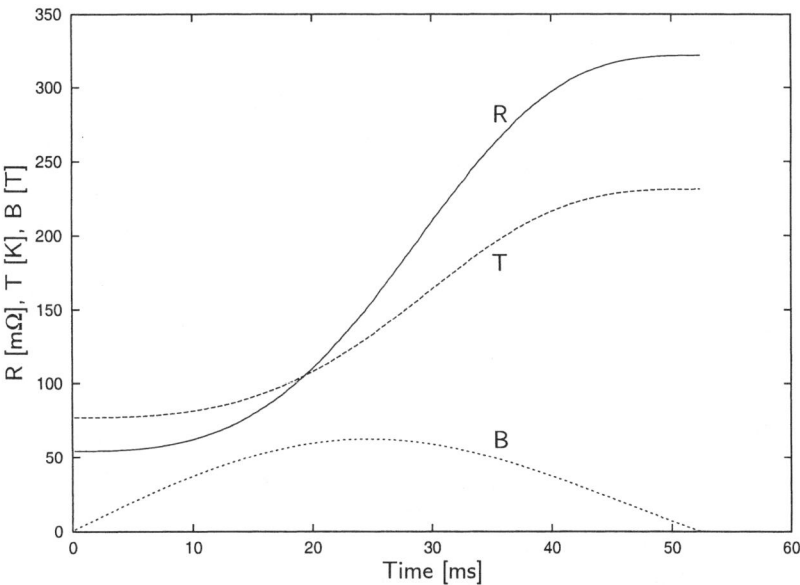

Fig. 4.3. Simulation of a capacitor C discharging into a coil (L, R) with temperature-dependent resistance R. The graph shows the temperature, the resistance and the central field of the coil as a function of time. Data for the circuit parameters and results can be found in Table 4.1

$$\dot{I}(x) = -\frac{R_{n+1}}{L}x + \frac{U_n}{L}.$$

We have assumed here a coil consisting of one conductor material only, which additionally has everywhere the same current density. Consequently, the conductor has everywhere the same temperature. In Fig. 4.3 the results of a simulation are shown. The circuit data and results of the simulation are summarized in Table 4.1. As the conductor material we have chosen copper. The simulation shows that after 24.7 ms a maximum field of 62.3 T is achieved; the length of the pulse is about 52.7 ms. After that time the coil resistance and temperature have risen from 54 mΩ to 322 mΩ at 77 K to 231 K, respectively. The estimation with constant circuit elements would give a field of 65.0 T, reached after 26.6 ms; the pulse length would be 54.3 ms.

Incorporation of the resistance increase is necessary for determining the evolution of the temperature in the coil. In the given example the varying resistance does not have much influence on the achievable field and the pulse length. This is due to the fact that at the time of the field maximum the temperature and resistance are about half-way between their respective starting and final values. The effect of temperature increase on the maximum field can therefore be expected to be bigger if the temperature span gets bigger. The conclusion is that one should not deliberately make the discharge process slow, since this would give the coil enough time to increase its resistance, which would mean a lower maximum current and therefore field. The idea is to achieve the maximum field at a time when the temperature of the coil has not risen too much.

Of course there is then another limiting factor. A faster discharge process for a fixed coil can be achieved only by a smaller capacitor, but this implies higher voltages when one wants to store the same amount of energy. Higher voltages then may cause problems with the insulation of the coil. Additionally, with a faster discharge the influence of eddy currents becomes more visible. This makes it not only difficult to achieve a fast discharge (yet smaller capacitor, yet higher voltages), but also the heating due to eddy currents may overcompensate what one wanted to achieve, namely a field maximum at a low temperature.

4.1.3 Magnetoresistance

Here we demonstrate how to incorporate the magnetoresistance of the conductor into the simulation. Basically we only need to modify the algorithm in Sect. 4.1.2 and calculate for each time step the resistance increase due to the ohmic heating and additionally due to the magnetoresistance, as was outlined in Sect. 1.3.4. Because the magnetic field changes over the coil cross-section, this means that the heating of the coil will become nonuniform and a temperature distribution evolves.

For the simulation we use here the model of a wire-wound coil consisting of N_r current rings in the radial direction and N_z current rings in the axial

Table 4.1. Data and results for a simulation of an RLC circuit consisting of a capacitor C and a coil, described by its inductance L and temperature-dependent resistance R. At the beginning of the discharge process the current is zero and the capacitor is charged to $U_0 = 15\,\mathrm{kV}$. The initial temperature T_i was chosen to be 77 K, and the conductor was copper wire with a cross-section of $A_q = 10\,\mathrm{mm}^2$. The simulation of the differential equation (4.1) with the temperature-dependent resistance was performed with a Runge-Kutta method. The results are a maximum field in the center of the coil of $B_{\max} = 62.3\,\mathrm{T}$, reached 24.7 ms after the start of the discharge process. The pulse length is $t_{\mathrm{pulse}} = 52.7\,\mathrm{ms}$. The development of temperature, resistance and central field as a function of time are shown in Fig. 4.3

Input data:			
inner radius	a_1	10	mm
outer radius	a_2	100	mm
coil height	h	200	mm
conductor	copper		
number of windings	$N = N_r\,N_z$	$720 = 18 \times 40$	
cross-section of wire	A_q	10	mm^2
capacitance	C	20	mF
load voltage	U_0	15	kV
inductance	L	15	mH
initial temperature	T_i	77	K
initial resistance	$R(T_i)$	54	mΩ
Simulation results:			
maximum field	$B_{\max}$	62.3	T
time of field maximum	$t_{\max}$	24.7	ms
final temperature	T_f	231	K
final resistance	$R(T_f)$	322	mΩ
pulse length	t_{pulse}	52.7	ms
peak reverse voltage	U_{rev}	-10.9	kV

direction. The set of rings shall form an even distribution of wires within the coil cross-section. For the estimation of the magnetoresistance we use only the magnetic field at the center of each wire, i.e. we neglect variations of the magnetic field within the cross-section of a single wire.

The results of the simulation are shown in Fig. 4.4, which also gives the results for a simulation of the same coil without magnetoresistance. The incorporation of magnetoresistance has almost no effect on the maximum field, but changes are visible for the coil resistance and temperature. The temperature in Fig. 4.4 is the highest temperature in the coil, and the distribution of the temperature over the $N_r \cdot N_z$ wires after the current pulse is shown

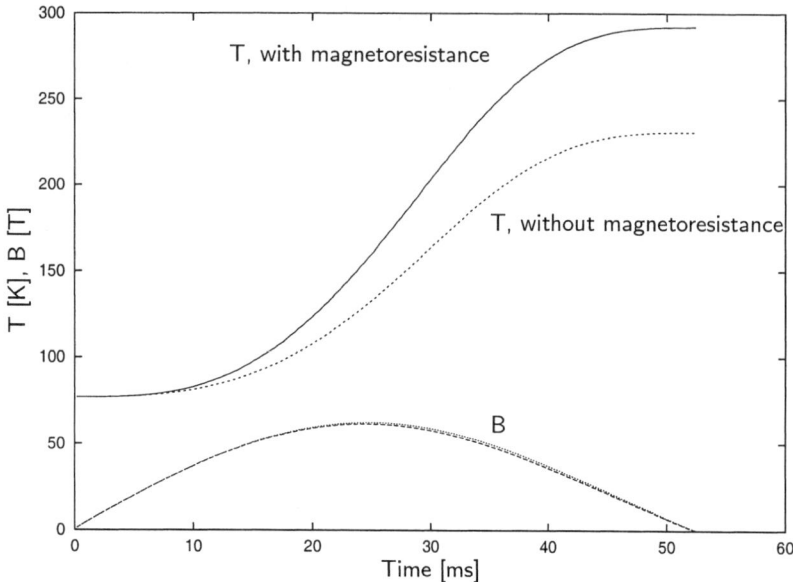

Fig. 4.4. Simulation of a capacitor discharge into a coil, where the resistance increase due to ohmic heating and due to magnetoresistance is taken into account. For comparison the simulation without the magnetoresistive effects is also shown. The differences in the generated field from both simulations are almost invisible; however, a clear change in the temperature is visible. In the case of the magnetoresistive simulation the temperature is that from the hottest spot in the cross-section. The temperature distribution at the end of the discharge process is shown in Fig. 4.5. Both simulations were performed for the same coil data (see Table 4.1)

in Fig. 4.5. In the simulation the highest temperature increase occurs in the midplane at the inner surface of the coil, where the magnetic field has its highest values and therefore the magnetoresistance has its highest impact on the temperature increase. The temperature falls towards the outer radius of the coil from 290 K to about 225 K; at the top and the bottom of the cross-section near to the outer surface a slight increase in temperature can be seen. This is caused by the radial field component, which has at these positions a maximum. For the magnetoresistance the transversal field is responsible, this means we have to calculate the axial and the radial field component at the center of each wire and use the absolute value of the magnetic field there.

Table 4.2 shows a comparison for the results of the three different ways of calculation presented in this chapter. The estimation with constant circuit elements is the fastest one. If used with the resistance value at the initial temperature, this method inherently overestimates the maximum field, the time of this field maximum and the pulse length. In the given example the overestimation is about 6% for the maximal field, roughly 9% for the time of the maximum and 3% for the pulse length. The main advantage of this

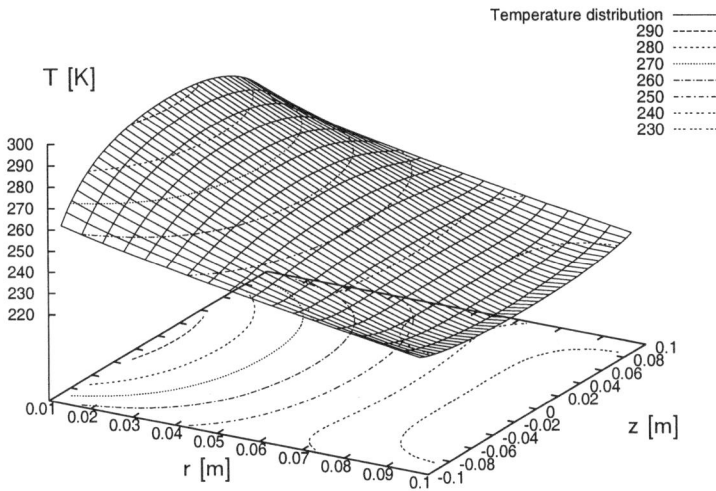

Fig. 4.5. Temperature distribution in the coil from Table 4.1 after the time t_{pulse}. The distribution is caused by the different magnetoresistance of the coil wire along its cross-section, which is caused by the variation of the magnetic field strength. The highest temperature increase occurs at the position of the highest field in the cross-section. The maximum temperature of about 290 K lies in the midplane at the inner surface of the coil; the lowest temperature can be found in the midplane very near to the outer radius of the coil. The field in the midplane at the outer radius has the opposite polarity to that of the central field. Hence, somewhere in the midplane there is a point where the field disappears; at this point there is no additional heating due to magnetoresistive effects

Table 4.2. Comparison of the main parameters of a RLC discharge process, such as the time and value of the field maximum ($t_{\max}$, $B_{\max}$), the pulse length t_{pulse} and the final temperature T_f. With increasing sophistication the methods of calculation treat the resistance as being constant ($R = \text{const}$), being a function of the coil temperature ($R = R(T)$) and finally depend on the temperature and the magnetic field ($R = R(T, B)$)

	$R = \text{const}$	$R = R(T)$	$R = R(T,B)$	
$B_{\max}$	65.0	62.3	61.5	T
$t_{\max}$	26.6	24.7	24.4	ms
t_{pulse}	54.3	52.7	52.7	ms
T_f		231	292	K

method is that it allows an overview of the circuit even without a computer; however, it lacks a temperature evolution. This is overcome by the numerical simulation of the ohmic heating of the wire, which determines the time and the value of the field maximum with an estimate that is about 1% too high. The next refinement is the incorporation of magnetoresistance, which causes a considerably stronger heating of the coil. Without magnetoresistance the final temperature of the coil would be underestimated by about 20%.

Of course, the effect of the magnetoresistance depends on the coil geometry, the conductor material and the allowed temperature span. The example given here only demonstrates that the effect usually has a very small influence on the generated field but leads to a non-negligible temperature increase. If this additional temperature increase is too high for the coil construction, then the pulse length must be decreased.

4.1.4 Eddy Currents

The trick of extending a coil in the axial direction to infinity is also useful for the calculation of eddy currents. The basic equation is the first Maxwell equation:

$$\frac{d}{dt} \int \boldsymbol{B} \cdot d\boldsymbol{F} = - \int \boldsymbol{E} \cdot d\boldsymbol{s} . \tag{4.10}$$

This equation states that a variation of the magnetic flux through a surface F causes an electric voltage $-\int \boldsymbol{E} \cdot d\boldsymbol{s}$ along the border of the surface F. This voltage gives rise to electrical currents in a conducting material; these currents are called eddy currents. The calculation of the exact distribution of the eddy currents can be a difficult mathematical and numerical problem, as the following small thought experiment will show. We put a massive piece of conductor suddenly into an external magnetic field. Then eddy currents will be induced in the conductor. Due to Lenz's rule the eddy currents will flow in such a direction as to oppose the external flux change. Hence the total field consists of the external field plus the field from the eddy currents. If we assume a normal and not a superconducting material, then these eddy currents will decay with a certain time constant and the external field will diffuse into the conductor. As soon as the diffusion process has reached an equilibrium, the total field is again equal to the external field (we assume the conductor to be non-ferromagnetic and neglect dia- or paramagnetic properties). The mathematical difficulties arise because it is the total flux which must be used in (4.10), that is the external flux and the flux from the eddy current distribution. One has to find a self-consistent solution. Furthermore, we have to deal with a time-dependent problem, since the eddy currents decay with a certain time constant [129–153].

Because the calculation of the diffusion of the magnetic field is difficult and on the other hand we want to make only an estimation, we neglect the diffusion altogether. This is valid if the change of the external field is slow compared to the characteristic diffusion time. Instead of dealing with diffusion times we can also speak of diffusion lengths, which means that the geometric dimensions of the conductor must be smaller than or comparable to the characteristic diffusion length. This length is also called the skin depth d. Its value depends on the resistivity ϱ and the permeability μ of the conductor and the rate (= frequency) with which the external field changes:

$$d = \sqrt{\frac{2\varrho}{\mu\omega}}. \tag{4.11}$$

For copper at room temperature ($\varrho = 1.7 \times 10^{-8}\,\Omega\,\text{m}$, $\mu = \mu_0 = 4\pi \times 10^{-7}\,\text{V\,s/A\,m}$) and a frequency of 50 Hz this length is $d = 9.3\,\text{mm}$. If the external field changes at such a rate and the conductor dimensions are less than about 9 mm, then diffusion effects can be neglected to a first order.

With this simplification we regard the external field as being identical with or without the conducting material; the eddy currents can then be calculated by considering only the external flux in (4.10):

$$\frac{\mathrm{d}}{\mathrm{d}t}\int \boldsymbol{B}_{\text{ext}}\cdot\mathrm{d}\boldsymbol{F} = -\int \boldsymbol{E}\cdot\mathrm{d}\boldsymbol{s}. \tag{4.12}$$

Now we can calculate the eddy currents for a thick-walled cylinder with a slit along its axis. The aim of the estimation is to find expressions for the inductance and the resistance of the eddy currents of such a structure. Having done this, the analysis of a coil with eddy currents is equivalent to the circuit of a transformer: in the primary loop lies the inductance and

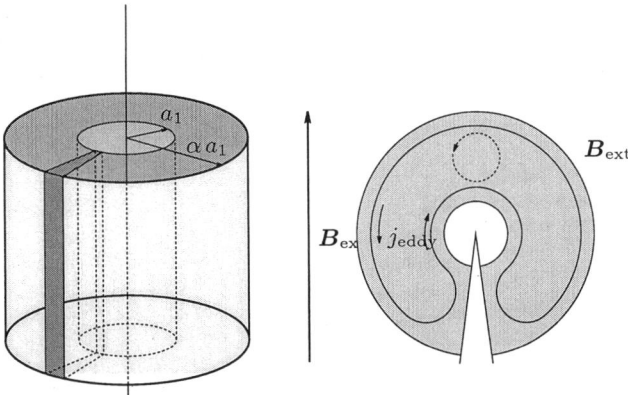

Fig. 4.6. A thick-walled cylinder with a slit subjected to a time-varying external field. The field has only an axial component. The lower picture shows the flow of the eddy currents; the pattern emerges from combining numerous smaller loops like the one indicated by the dashed line style. The radial current components of two neighboring loops cancel each other and only an azimuthal component remains. As a result the currents flow on the inner and the outer surfaces of the cylinder in opposite directions. For the estimation performed here we treat the cylinder as infinitely long and neglect diffusion effects of the magnetic field into the cylinder

142   4. Pulsed Field Facilities

resistance of the coil itself (L, R); the secondary, closed loop consists of the inductance and resistance of the eddy current circuit $(L_{\text{eddy}}, R_{\text{eddy}})$. Between both inductances there exists a mutual inductance.

Thick-walled Cylinder with Slit. The geometry of the thick-walled cylinder with slit is shown in Fig. 4.6. Opposed to an axial time-varying field, eddy currents flow in opposite directions on the inner and the outer side of the cylinder. The distribution of the eddy current density is found by regarding the small, dashed path of the eddy currents in Fig. 4.6. Two neighboring such paths cancel their radial current components and only an azimuthal component of the current density remains. With the center of the dashed loop lying half-way between the inner and outer radii of the cylinder we find from the Maxwell's equation (4.10):

$$E\, 2\pi y = -\dot{B}_{\text{ext}}\, y^2 \pi \,, \tag{4.13}$$

where y is the distance to the center of the loop. For the distribution of the eddy current density in the cylinder this means that it is linear in the radius r and disappears at $r/a_1 = (\alpha + 1)/2$:

$$j_{\text{eddy}}(x) = -\frac{a_1}{2\varrho}\, \dot{B}_{\text{ext}}\, \frac{\alpha + 1 - 2x}{2} \quad \text{with} \quad 1 \leq x \leq \alpha\,. \tag{4.14}$$

Again, we have assumed the cylinder to be infinitly long and have neglected the diffusion effects of the magnetic field into the cylinder. Performing an integration we find for the total eddy current:

$$I_{\text{eddy}} = h\, a_1 \int_1^{\frac{\alpha+1}{2}} j_{\text{eddy}}\, dx = h\, a_1 \left(-\frac{a_1}{2\varrho}\dot{B}_{\text{ext}}\right)\frac{(\alpha-1)^2}{8}\,. \tag{4.15}$$

Because the current flows in the cylinder back and forth we must integrate only from the inner surface up to the radius $a_1 \frac{\alpha+1}{2}$. The height of the cross-section is h. The ohmic power is found by an integration of $\varrho\, j_{\text{eddy}}^2$ over the total cylinder volume with height h:

$$P = h \int_1^\alpha \varrho\, j_{\text{eddy}}^2\, 2\pi a_1 x\, a_1\, dx = \varrho \left(-\frac{a_1}{2\varrho}\dot{B}_{\text{ext}}\right)^2 h\, \pi a_1^2\, \frac{(\alpha-1)^3(\alpha+1)}{12}\,. \tag{4.16}$$

The magnetic field is calculated as

$$B_{\text{eddy}}(x) = -\mu_0 \left(\frac{a_1}{2\varrho}\dot{B}_{\text{ext}}\right) \frac{a_1}{2}\,(\alpha^2 - x)(1 - x)\,, \tag{4.17}$$

and it disappears outside of the cylinder. The magnetic energy associated with this field distribution is hence

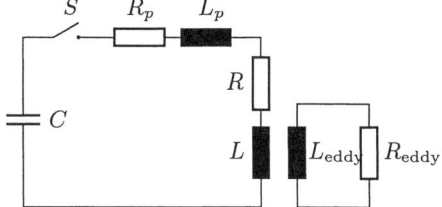

Fig. 4.7. Circuit for the simulation of eddy currents in a coil. After closing the switch S the capacitor discharges into the protection coil (L_p, R_p) and the high field coil (L, R). The eddy currents in the high-field coil define a secondary loop, consisting of a resistance R_{eddy} and an inductance L_{eddy}, which is closely coupled to the high-field coil; as the coupling factor we assume $k = 0.9$

$$W = \frac{1}{2\mu_0}\left(\mu_0 \left[\frac{a_1}{2\varrho}\dot{B}_{\text{ext}}\right]a_1\right)^2 h\,\pi a_1^2\,\frac{(\alpha+1)(\alpha-1)^3}{30}. \tag{4.18}$$

Now we get for the resistance and the inductance:

$$R_{\text{eddy}} = \varrho\,\frac{2\pi}{h}\,\frac{8}{3}\,\frac{\alpha+1}{\alpha-1}, \tag{4.19}$$

$$L_{\text{eddy}} = \mu_0\,\frac{\pi a_1^2}{h}\,\frac{8}{15}(\alpha^2-1), \tag{4.20}$$

and we arrive at a time constant of

$$\tau_{\text{eddy}} = \frac{L_{\text{eddy}}}{R_{\text{eddy}}} = \frac{\mu_0}{\varrho}\,a_1^2\,\frac{(\alpha-1)^2}{10}. \tag{4.21}$$

Eddy Currents in a Coil. In this section we incorporate eddy current effects in the high-field coil. We use the equivalent circuit as outlined above and perform a simulation of a real coil system. The circuit is shown in Fig. 4.7. It consists of a capacitor C, which can be discharged by closing a switch S into a protection coil (L_p, R_p) and a high-field coil (L, R). The eddy currents in the coil are simulated as a loop comprised by an inductance L_{eddy} and a resistance R_{eddy}. The high-field coil L and the eddy current coil L_{eddy} are strongly coupled. The data for all the circuit elements are given in Table 4.3.

The circuit parameters have values allowing a rather high frequency of the damped oscillation, which means that we can neglect any effects of the ohmic heating in the high-field coil, i.e. all resistors are treated as being constant. At a temperature of 77 K the high-field coil has a resistance of $R = 44\,\text{m}\Omega$, the inductance of the wire-wound coil is $L = 1.8\,\text{mH}$. The wire has a cross-section of $2.2 \times 3.2\,\text{mm}^2$, and its resistivity at 77 K is $\varrho = 5.5 \times 10^{-9}\,\Omega\,\text{m}$.

The eddy currents in the wire have to run back and forth, which gives half the cross-section and twice the length, so that a first rough estimation would set the resistance of the eddy currents to be four times the resistance of the coil. A more elaborate estimation takes account of the nonuniform

Table 4.3. Circuit parameters for Fig. 4.7. The data for the capacitor, the protection coil and the high-field coil are known a priori, and the inductance and resistance of the eddy current loop are estimated as outlined in the text

Parameter		Value	
capacitor	C	5	mF
protection coil	R_p	116	mΩ
	L_p	1.33	mH
high-field coil	R	44	mΩ
	L	1.80	mH
eddy-current loop	R_{eddy}	235	mΩ
	L_{eddy}	0.026	mH
coupling factor	k	0.9	

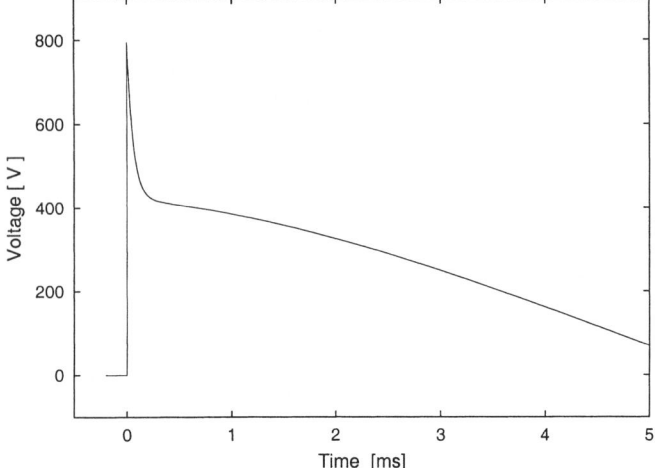

Fig. 4.8. Numerical simulation of the circuit from Fig. 4.7. Shown is the voltage across the inductance L_p as a function of time. This voltage is proportional to the derivative of the current through the high-field coil. Measuring the magnetic field with an induction coil would therefore result in a similar behavior. The sharp peak at the beginning of the simulation is caused by the eddy-current loop, it reflects also the time constant of this loop of $\tau_{eddy} = 0.11$ ms. The slower process is due to the damped oscillation of the primary loop, comprising the series connection $C - R_p - L_p - R - L$ (see Fig. 4.7)

distribution of the eddy currents in the wire. The resistance of one turn of wire with mean radius r and cross-section A_q is

$$R_{1\text{turn}} = \varrho \, \frac{2\pi r}{A_q} \, . \tag{4.22}$$

From (4.19) we find in the limit of a very thin cylinder the resistance of the wire to be

$$R_{\text{eddy}} = \frac{16}{3} R_{1\text{turn}}. \tag{4.23}$$

So we estimate the resistance of the eddy current loop to be $R_{\text{eddy}} = \frac{16}{3} \cdot 44\,\text{m}\Omega \approx 235\,\text{m}\Omega$. Using the same model we find for the time constant of the one turn of wire (see (4.21))

$$\tau_{\text{eddy}} = \frac{\mu_0}{\varrho} \frac{d^2}{10}, \tag{4.24}$$

and with a thickness of the wire of $d = 2.2\,\text{mm}$ we find for the time constant of the eddy current loop that $\tau_{\text{eddy}} = 0.11\,\text{ms}$ and the inductance follows as $L_{\text{eddy}} = 0.026\,\text{ms}$. The coupling factor between L and L_{eddy} is taken as $k = 0.9$.

With all data now determined a simulation of the circuit was performed, and the result is given in Fig. 4.8. It shows the voltage across the inductance L_p of the protection coil as a function of time. The voltage across L_p is proportional to the derivative of the current through the high-field coil. An induction coil in the bore of the high-field coil, which can be used for measuring the magnetic field, shows the same behavior. The outstanding feature in Fig. 4.8 is the sharp peak just after the discharge process of the capacitor has been started. It is indeed the eddy current loop which causes this peak, as simulations without the loop clearly show.

The same behavior can be found in a real measurement, as shown in Fig. 4.9, where the voltage from an induction coil, which was put into the bore of the high-field coil, is drawn as a function of time. The peak at $t = 0$ is not as high as in the simulation, but the decay of the first peak happens within roughly the same time. The differences in the simulation and the actual measurement may be due to a still too coarse estimation of the parameters of the eddy-current loop. Further simulations with other combinations for the resistance, inductance and coupling factor of the eddy current loop show that the height of the peak is sensitive to such variations.

4.1.5 RLC Circuit with Crowbar Diode

A very important variation of the simple RLC circuit consists in the addition of a diode parallel to the capacitor. Usually the diode is in series with a resistor (see Fig. 4.10), and the modification is often referred to as a RLC circuit with crowbar diode. Initially the diode sees a reverse voltage, and the discharge of the capacitor C into the coil starts as presented in Sect. 4.1.2. However, shortly after the current maximum the diode sees a voltage in the forward direction and starts to conduct. The current and hence the field in the coil then decay exponentially with a time constant of

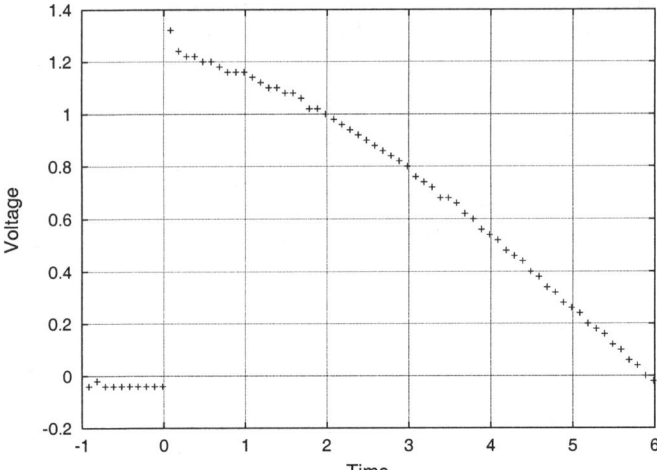

Fig. 4.9. Real measurement with an induction coil of the circuit from Fig. 4.7. The principal behavior is identical to the numerical simulation from Fig. 4.8. The length of the long-term oscillation and the decay of the initial peak are almost identical; only the height of the first peak is smaller in the real measurement than in the simulation

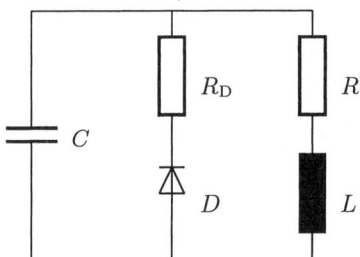

Fig. 4.10. Principal circuit of a capacitor C discharged into a coil (resistance R and inductance L), extended with a so-called crowbar diode D parallel to the capacitor. The crowbar diode is in series with a crowbar resistor R_D

$$\tau_{\text{decay}} = \frac{L}{R + R_D}, \qquad (4.25)$$

The magnitude of the crowbar resistance R_D determines for a fixed coil the decay rate; furthermore the reverse voltage on the capacitor C depends on the magnitude of the crowbar resistance. As in to Sect. 4.1.2 we present a simulation with a temperature-dependent coil resistance R; all the other circuit elements are treated as temperature-independent and ideal elements. The numerical simulation is similar to the previous RLC simulation. Additionally, the voltage across the diode is monitored, and as soon as this voltage becomes negative the crowbar diode and resistor branch is put into the simulation.

In Fig. 4.11 the results of a simulation are shown. The circuit data and results of the simulation are summarized in Table 4.4. As the conductor material we have chosen copper. The simulation shows that after 24.7 ms a maximum field of 62.3 T is achieved. When the crowbar branch starts to conduct, the current and field in the coil decay exponentially. Therefore, one cannot talk about a pulse length anymore. A 'pulse length' might be defined by the time by when the field has decayed to 90%, but that is rather arbitrary. Following the simulation for sufficient time we can determine the final values of the coil resistance and temperature. We find a resistance increase from 54 mΩ to 394 mΩ and a temperature increase from 77 K to 275 K.

4.2 Inductive Energy Transfer

4.2.1 Ideal Coils with No Coupling

Here we investigate the features of inductive energy transfer: a primary current source (battery, generator) charges a storage inductance, then the stored energy is transferred with a certain efficiency to a load inductance, namely a high-field coil. Since for pulsed magnet systems the produced field is proportional to the square root of the magnetic energy in the high-field coil, we

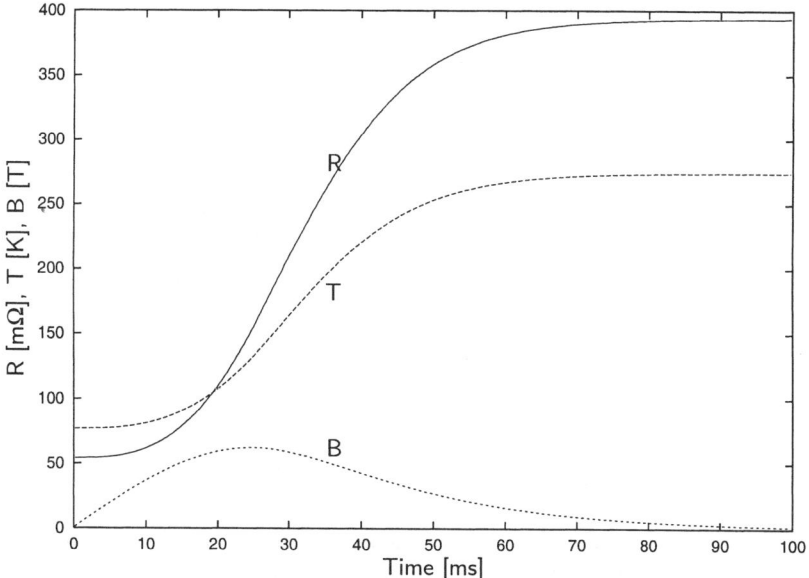

Fig. 4.11. Simulation of a capacitor C discharging into a coil (L, R) with temperature-dependent resistance R. The graph shows the temperature, the resistance and the central field of the coil as a function of time. The data for the circuit parameters and results can be found in Table 4.4

Table 4.4. Data and results for a simulation of a RLC circuit with crowbar diode. All circuit elements except the coil resistance R are considered as constant. The discharge process begins with zero current in the capacitor charged to $U_0 = 15\,\text{kV}$. The initial temperature T_i was chosen to be 77 K, and the conductor was copper wire with a cross-section of $A_q = 10\,\text{mm}^2$. The results are a maximum field in the center of the coil of $B_{\max} = 62.3\,\text{T}$, reached 24.7 ms after the start of the discharge process. Temperature, resistance and central field as a function of time are shown in Fig. 4.11

Input data:			
inner radius	a_1	10	mm
outer radius	a_2	100	mm
coil height	h	200	mm
conductor	copper		
number of windings	$N = N_r\,N_z$	$720 = 18 \times 40$	
cross-section of wire	A_q	10	mm^2
capacitance	C	20	mF
load voltage	U_0	15	kV
inductance	L	15	mH
initial temperature	T_i	77	K
crowbar resistance	$R(T_i)$	300	mΩ
Simulation results:			
maximum field	$B_{\max}$	62.3	T
time of field maximum	$t_{\max}$	24.7	ms
final temperature	T_f	275	K
final resistance	$R(T_f)$	394	mΩ
peak reverse voltage	U_{rev}	-3.3	kV

wish to find a circuit which transfers as much energy as possible. Because the governing equations can be quite long we first concentrate on the ideal circuit shown in Fig. 4.12. Two ideal coils and a resistor are connected in parallel. By an ideal coil we mean a coil having only inductance but no resistance. Technically such a coil may be realized by a superconducting coil. The resistor may represent a fuse or a switch, whose initial resistance is zero, and current flows only in the loop I: $I_1 = I_3 = I_0$ and $I_2 = 0$. Within a short time interval the resistance is increased to infinity, thereby transferring the energy: $I_3 = 0$ now and $I_1 = I_2$. The efficiency can be immediately derived from the conservation of flux.[1] For the initial stored energy we write

---

[1] As mentioned above, an ideal coil might be realized by a superconducting coil, which then prevents magnetic flux lines from leaving a closed loop of conductor,

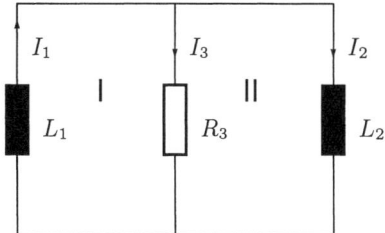

Fig. 4.12. Inductive energy transfer with storage inductance L_1, load inductance L_2 and switch resistor R_3

$$W_0 = \frac{1}{2} L_1 I_0^2 . \tag{4.26}$$

After the energy is transferred, we have in both inductances the same current I flowing. Hence the conservation of the magnetic flux Φ is written as

$$\Phi = \text{const.} \Rightarrow \tag{4.27}$$

$$L_1 I_0 = L_1 I + L_2 I . \tag{4.28}$$

The energy transferred into the second inductance L_2 can now immediately be written as

$$W_2 = \frac{L_1 L_2}{(L_1 + L_2)^2} W_0 . \tag{4.29}$$

The transferred energy W_2 is a maximum, if $L_1 = L_2$. In this case the final energies in both inductances are

$$W_1 = W_2 = \frac{1}{4} W_0 , \tag{4.30}$$

and after the transfer 25% of the initial energy remains in the inductance L_1 as well as in L_2 and the other 50% is dumped into the resistor. Now we derive equations for the currents as they change with time. Applying Kirchhoff's rules in Fig. 4.12 we get the following three equations:

$$I_1 - I_2 - I_3 = 0 , \tag{4.31}$$

$$L_1 \dot{I}_1 + R_3 I_3 = 0 , \tag{4.32}$$

$$L_2 \dot{I}_2 - R_3 I_3 = 0 . \tag{4.33}$$

The initial conditions are $I_1(0) = I_0$ and $I_2(0) = 0$, which gives the following solution:

$$I_1(t) = I_\infty \left(1 + \frac{L_2}{L_1} e^{-(\lambda_1 + \lambda_2) t} \right) , \tag{4.34}$$

$$I_2(t) = I_\infty \left(1 - e^{-(\lambda_1 + \lambda_2) t} \right) , \tag{4.35}$$

---

e.g. a coil. Furthermore, we also assume the quasi-stationary case, which excludes energy losses due to electromagnetic waves emanating from the circuit.

with use of the abbreviations

$$\lambda_1 = \frac{R_3}{L_1}, \quad \lambda_2 = \frac{R_3}{L_2}, \quad I_\infty = \frac{L_1}{L_1 + L_2} I_0 . \tag{4.36}$$

The current in the storage inductance decreases monotonically from its starting value I_0 to the final value I_∞, and the current in the second inductance, the load coil, increases monotonically from zero to the same asymptotic value I_∞. The stored energy in the load coil L_2 as a function of time is written as

$$W_2(t) = W_0 \frac{L_1 L_2}{(L_1 + L_2)^2} \left(1 - e^{-(\lambda_1 + \lambda_2)t}\right)^2 . \tag{4.37}$$

In the asymptotic limit for $t \to \infty$ this energy is a maximum if $L_1 = L_2$ and then has the value $\frac{1}{4} W_0$. The commutating resistor R_3 does not affect the final energy; however, it does influence the rise time of the current. The circuit time constant is

$$\tau = \frac{1}{\lambda_1 + \lambda_2} = \frac{1}{R_3} \frac{L_1 L_2}{L_1 + L_2} . \tag{4.38}$$

The time for the energy $W_2(t)$ to reach, for instance, 90% of its final value is the smaller, the higher the resistor R_3 is. From this viewpoint the ideal resistor should be as high as possible to achieve a fast energy transfer. Unfortunately, there occur high voltages across the resistor ($U_3 = R_3 I_0$ at $t = 0$), so that in a real system an upper limit for R_3 exists.

In the circuit analyzed until now we have assumed that the two coils are far apart so that they impose no influence on each other. A non-zero mutual inductance changes the energy transfer remarkably.

4.2.2 Ideal Coils with Coupling

The analysis of the new circuit, shown in Fig. 4.13, is similar to the one before. Two ideal inductances with mutual inductance M are connected parallel to a resistor. At the beginning of the energy transfer the current in the coil L_1 is I_0 and zero in the coil L_2. With the help of Kirchhoff's rules we derive:

$$I_1 - I_2 - I_3 = 0 , \tag{4.39}$$
$$L_1 \dot{I}_1 + M \dot{I}_2 + R_3 I_3 = 0 , \tag{4.40}$$
$$L_2 \dot{I}_2 + M \dot{I}_1 - R_3 I_3 = 0 . \tag{4.41}$$

The solution for this system of differential equations is:

$$I_1(t) = I_\infty + (I_0 - I_\infty) e^{-\chi t} , \tag{4.42}$$
$$I_2(t) = I_\infty \left(1 - e^{-\chi t}\right) . \tag{4.43}$$

With the introduction of the following constants

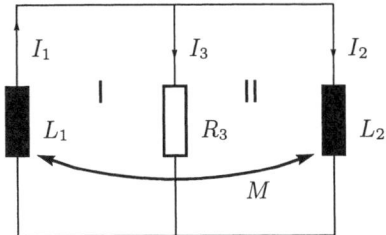

Fig. 4.13. Inductive energy transfer with storage inductance L_1, load inductance L_2 and switch resistor R_3. The two inductances are coupled by a mutual inductance M

$$k = \frac{M}{\sqrt{L_1 L_2}}, \quad \lambda_1 = \frac{R_3}{L_1}, \quad \lambda_2 = \frac{R_3}{L_2}, \quad x = \frac{L_2}{L_1} = \frac{\lambda_1}{\lambda_2} \quad (4.44)$$

we find

$$\chi = \frac{1}{1 - k^2} \left(\lambda_1 + \lambda_2 + 2k \sqrt{\lambda_1 \lambda_2} \right), \quad (4.45)$$

$$I_\infty = \frac{\chi - \lambda_1}{\chi} \frac{1}{1 + k \sqrt{\frac{\lambda_1}{\lambda_2}}} I_0. \quad (4.46)$$

Now we can take a look at the transferred energies. For simplicity we express all energies in units of the initial energy W_0 and the inductance L_2 in units of L_1 ($L_2 = x L_1$). In the asymptotic limit $t \to \infty$ the energies in the two coils are

$$w_1 = \frac{W_1}{W_0} = \left(\frac{1 + k\sqrt{x}}{1 + x + 2k\sqrt{x}} \right)^2, \quad (4.47)$$

$$w_2 = \frac{W_2}{W_0} = \left(\frac{1 + k\sqrt{x}}{1 + x + 2k\sqrt{x}} \right)^2 x. \quad (4.48)$$

Linked with the mutual inductance is the energy

$$w_m = \frac{W_m}{W_0} = \left(\frac{1 + k\sqrt{x}}{1 + x + 2k\sqrt{x}} \right)^2 2k\sqrt{x} \quad (4.49)$$

and the losses in the resistor R_3 are

$$w_R = 1 - w_1 - w_2 - w_M = \frac{(1 - k^2) x}{1 + x + 2k\sqrt{x}}. \quad (4.50)$$

All these energies depend only on the coupling constant k and the ratio of the self inductances $x = \frac{L_2}{L_1}$. To achieve a high efficiency of the energy transfer, one should use a combination of k and x for which w_2 is as high as possible. The allowed values are $0 \leq k \leq 1$ and $0 \leq x$.

For large values of x the energy in the inductance L_2 is written as

Table 4.5. Relative energies stored in the two inductances (w_1, w_2), the mutual inductance (w_M) and dumped into the resistor (w_R), for a few selected values of the coupling constant k and the inductance ratio $x = \frac{L_2}{L_1}$ for an ideal coupled-coil system

	$w_1(k,x)$	$w_2(k,x)$	$w_M(k,x)$	$w_R(k,x)$
$k=0$	$\dfrac{1}{(1+x)^2}$	$\dfrac{x}{(1+x)^2}$	0	$\dfrac{x}{(1+x)}$
$k=1$	$\dfrac{1}{(1+\sqrt{x})^2}$	$\dfrac{x}{(1+\sqrt{x})^2}$	$\dfrac{2\sqrt{x}}{(1+\sqrt{x})}$	0
$\lim x \to \infty$	0	k^2	0	$1-k^2$
$x=1$	$\dfrac{1}{4}$	$\dfrac{1}{4}$	$\dfrac{1}{2}k$	$\dfrac{1}{2}(1-k)$

$$w_2(x,k) \to k^2 \quad \text{for} \quad x \to \infty. \tag{4.51}$$

This means that in the case of a totally coupled coil system and for $L_2 \gg L_1$ all the energy initially stored in L_1 is transferred into L_2 without any losses.

For a finite inductance ratio $x = \frac{L_2}{L_1}$ and a coupling constant of $k = 1$, (4.50) gives $w_R = 0$, and there are no resistive losses; however, the initial energy is not totally transferred to L_2. A part of the energy is attached to the storage inductance and to the mutual inductance. A coupling factor of $k = 1$ obviously means that the coupled-coil system acts as a transformer.

The results for the energies (4.47–4.50) for a few selected values of k and x are summarized in Table 4.5. Figure 4.14 shows the energy associated with the inductance L_2 as a function of the inductance ratio $x = L_2/L_1$ and the coupling constant k. The isoline of $w_2(k,x) = 0.25$ separates the x-k plane into three regions: for $x < 1$ the transferred energy is lower than 0.25; for $x > 1$ the isoline separates a region with energies $w_2(k,x) > 0.25$ above and a region with $w_2(k,x) < 0.25$ below the isoline. For a given coupling constant k there exists an optimal inductance ratio x, which is greater than 1, i.e. $L_2 \geq L_1$. The best x value can be calculated to be

$$x_{\max} = \left(\frac{k + \sqrt{1-k^2}}{1-2k^2}\right)^2 \quad \text{for} \quad k < \frac{1}{\sqrt{2}}. \tag{4.52}$$

The transferred magnetic energy in this case is

$$w_{2,\max} = \frac{1}{4}\frac{1}{1-k^2}. \tag{4.53}$$

For a coupling constant $k > \frac{1}{\sqrt{2}}$ there is no local maximum, but a global maximum for $x \to \infty$ with a value of k^2.

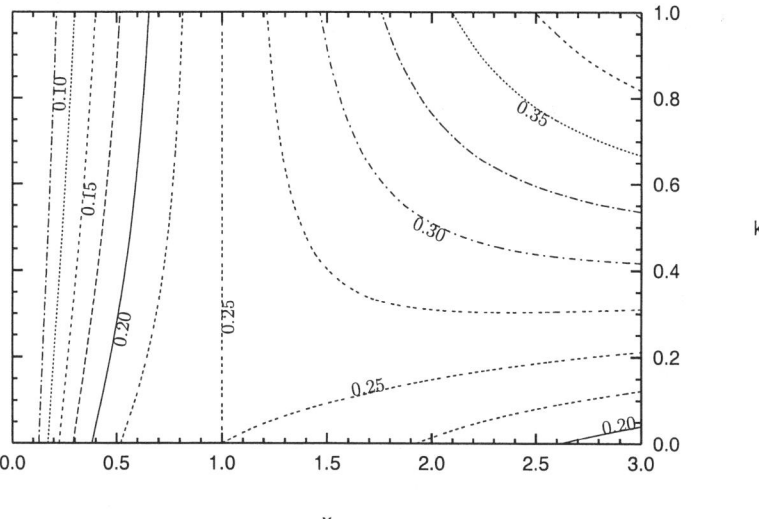

Fig. 4.14. Relative energy $w_2(k, x)$ associated with the inductance L_2 as a function of the inductance ratio $x = \frac{L_2}{L_1}$ and the coupling constant k

As for the uncoupled case the final energies are independent of the commutating resistor R_3. Only the rise and fall times are influenced, the circuit time constant being

$$\tau = \frac{1}{\chi} = \frac{1}{R_3} \frac{1-k^2}{\frac{1}{L_1} + \frac{1}{L_2} + \frac{2k}{\sqrt{L_1 L_2}}} \ . \tag{4.54}$$

The overvoltage across the resistor has a maximum value of $U_{\max} = R_3 I_0$ and has to stay below a certain level to guarantee safe operation of the energy transfer. Therefore, there exists an upper limit for the resistor R_3.

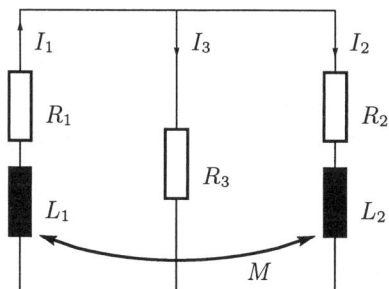

Fig. 4.15. Inductive energy transfer with storage inductance L_1, load inductance L_2 and switch resistor R_3. The two inductances are coupled with a mutual inductance M and have resistances R_1 and R_2, respectively

4.2.3 Real Coils with Coupling

Finally, the step is made from ideal coils to real coils: two coils with inductances L_1 and L_2 and a commutating resistor R_3 are connected in parallel. Between both coils there is some mutual inductance M and both have a finite resistance R_1 and R_2, respectively. Because there occur additional losses in the resistors R_1 and R_2, the efficiency of the energy transfer has to be lower than in the case of the coupled ideal coils from the last chapter. The circuit is shown in Fig. 4.15. The inductances as well as the resistors are regarded as constant. This is a simplification, especially since the resistance of the high-field coil is strongly dependent on the temperature, which rises in a typical experiment from 77 K to 350 K. One selects a combination of the circuit parameters that results in a maximum of the current through L_2, and therefore of the field too, occurring at a time $t_{\max}$. If this time is chosen to be sufficiently small, then the temperature increase at the time of the current maximum can be neglected and the model of constant circuit parameters gives reasonable results. With the given current directions of Fig. 4.15 we get from Kirchhoff's rules the following three equations:

$$I_1 - I_2 - I_3 = 0, \tag{4.55}$$

$$L_1\dot{I}_1 + M\dot{I}_2 + R_1 I_1 + R_3 I_3 = 0, \tag{4.56}$$

$$L_2\dot{I}_2 + M\dot{I}_1 + R_2 I_2 - R_3 I_3 = 0. \tag{4.57}$$

With the initial conditions $I_1(t=0) = I_0$ and $I_2(t=0) = 0$ the solution for this system is:

$$I_1(t) = I_{10}\, e^{-\chi_- t} + (I_0 - I_{10})\, e^{-\chi_+ t}, \tag{4.58}$$

$$I_2(t) = I_{20}\left(e^{-\chi_- t} - e^{-\chi_+ t}\right), \tag{4.59}$$

with

$$\chi_\pm = \frac{(\Lambda_1 + \Lambda_2 + 2k\sqrt{\lambda_1\lambda_2})}{2(1-k^2)} \left\{ 1 \pm \sqrt{1 - \frac{4(1-k^2)[\Lambda_1\Lambda_2 - \lambda_1\lambda_2]}{[\Lambda_1 + \Lambda_2 + 2k\sqrt{\lambda_1\lambda_2}]^2}} \right\}, \tag{4.60}$$

$$I_{10} = \frac{(\chi_+ - \Lambda_1)}{(\chi_+ - \chi_-)} \frac{\left(1 + k\frac{1}{\sqrt{\lambda_1\lambda_2}}\chi_-\right)}{\left(1 + k\frac{1}{\sqrt{\lambda_1\lambda_2}}\Lambda_1\right)} I_0, \tag{4.61}$$

$$I_{20} = \frac{(\chi_+ - \Lambda_1)}{(\chi_+ - \chi_-)} \frac{(\Lambda_1 - \chi_-)}{\lambda_1} \frac{1}{\left(1 + k\frac{1}{\sqrt{\lambda_1\lambda_2}}\Lambda_1\right)} I_0, \tag{4.62}$$

where we have used the abbreviations

$$\lambda_1 = \frac{R_3}{L_1}, \lambda_2 = \frac{R_3}{L_2}, \tag{4.63}$$

$$\Lambda_1 = \frac{R_1 + R_3}{L_1} = \frac{1}{\tau_1} + \lambda_1, \Lambda_2 = \frac{R_2 + R_3}{L_2} = \frac{1}{\tau_2} + \lambda_2. \tag{4.64}$$

In (4.64) we have introduced the two characteristic time constants of the coils. The current maximum occurs at a time

$$t_{\max} = \frac{1}{\chi_+ - \chi_-} \ln \frac{\chi_+}{\chi_-}. \tag{4.65}$$

The transferred magnetic energy at this time depends on six parameters: $w_2 = w_2(L_1, R_1, L_2, R_2, R_3, M)$. Now a combination of parameters has to be found which maximizes w_2 and through that the generated field, too. Using the coil time constants and the coupling constant $k = \frac{M}{\sqrt{L_1 L_2}}$ we can rewrite

$$w_2 = w_2(\tau_1, L_1, \tau_2, L_2, R_3, k). \tag{4.66}$$

In the limit of $\tau_1, \tau_2 \to \infty$ the results of the coupled ideal coils are reproduced, and w_2 depends only on L_1, L_2 and k. If both time constants are at least sufficiently big, then we can assume that the dependence of the transferred energy on the resistor R_3 is only weak. The used boundary conditions of tolerable current and voltage determine the resistor, and to a first order $w_2 = w_2(\tau_1, L_1, \tau_2, L_2, k)$.

The construction of an inductive energy transfer system for a given storage coil with time constant τ_1 and inductance L_1 now requires us to calculate the optimal values of time constant and inductance of the second (high-field) coil, τ_2 and L_2. The coupling constant k between the two coils is usually small. For a high efficiency one should also try to make the time constants τ_1 and τ_2 as big a possible. The maximum allowed current and voltage define the commutating resistor R_3 and, with the required level of initially stored energy, the inductance L_1.

4.2.4 Improvements

Besides increasing the coupling between the storage coil and the high-field coil we may ask whether there are no other methods for improving the efficiency of the inductive energy transfer. A hint of the answer comes from flux compression techniques. The basic circuit is given in Fig. 4.16, which shows a storage coil L_1 with variable inductance connected to a load inductance L_2. With an initial current I_i, the magnetic flux and energy are

$$\Phi = (L_1 + L_2) I_i, \tag{4.67}$$

$$W_i = \frac{1}{2}(L_1 + L_2) I_i^2 = \frac{1}{2} \Phi I_i. \tag{4.68}$$

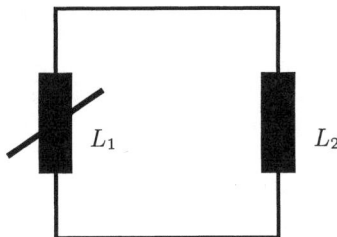

Fig. 4.16. Schematic circuit of flux compression. A variable storage inductance L_1 is connected to a load inductance L_2. Keeping the total flux constant, L_1 is reduced to zero, a process for which work is necessary. After the compression the magnetic flux is linked solely to the load L_2

Under the condition of flux conservation the storage inductance is now set equal to zero, $L_1 \to 0$. The final current is given by

$$\Phi = L_2 I_\mathrm{f} \tag{4.69}$$

and the final energy, which is now stored solely in the load coil L_2, is written as

$$W_\mathrm{f} = \left(1 + \frac{L_2}{L_1}\right) W_\mathrm{i} > W_\mathrm{i} . \tag{4.70}$$

The final energy is greater than the initial energy, which implies that there is work necessary in order to make L_1 equal to zero. Depending on the origin of this work we distinguish between electromagnetic or explosive flux compression.

Advanced inductive energy transfer circuits now do not resort to drastic explosions but use the principle of varying the storage inductance. The storage coil consists of several sections, which are either stepwise switched parallel or are stepwise switched out.

Parallel Switching Scheme. The principle of the parallel switching scheme is shown in Fig. 4.17. The current from the storage coil L_1 is initially short-circuited by the switch S_0, and so no current flows through the load coil L_2. By opening the switch the current is forced to flow through the load coil, and energy from the storage coil is transferred to the load coil. The switch has to be specially designed because there occur high inductive voltages over it during the opening process. As soon as the energy has been transferred, the two sections L_{11} and L_{12} of the storage coil are switched from the serial into the parallel condition. This is done by closing the two switches S_1 and S_2 (see the right-hand part of Fig. 4.17). The excessive current is short-circuited via the switch S_3, which is finally opened, thereby transferring again energy from the storage coil into the load. Basically this is the same circuit as in the

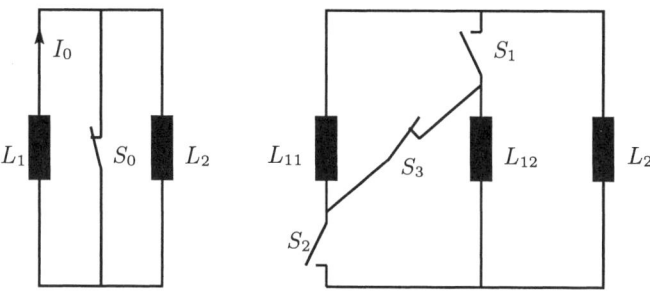

Fig. 4.17. Parallel switching scheme. The first step is shown on the left-hand side. By opening the switch S_0, energy from the storage coil L_1 is transferred into the load coil L_2. In the second step (right-hand side) the two parts of the storage coil, L_{11} and L_{12}, are switched from the initial serial into the parallel state by closing S_1 and S_2. Opening the switch S_3 transfers energy again from the storage system into the load

left-hand part of Fig. 4.17 with L_1 being the parallel-connected L_{11} and L_{12}, and S_0 identified by the switch S_3.

If two inductances L_m and L_n with mutual inductance $M = k\sqrt{L_m L_n}$ are connected in series or parallel then the resulting inductances are

$$L_{\text{ser}} = L_m + L_m + 2k\sqrt{L_m + L_n} \quad \text{for the serial case,} \tag{4.71}$$

$$L_{\text{par}} = \frac{L_m L_n}{L_m + L_n} \frac{L_m + L_n + 2k\sqrt{L_m L_n}}{L_m + L_n} \quad \text{for the parallel case.} \tag{4.72}$$

A few special cases are summarized in Table 4.6. If we allow only positive coupling between the two inductances, $0 \leq k \leq 1$, then the inductance in the serial case is always greater than in the parallel case:

$$L_{\text{ser}} > L_{\text{par}} . \tag{4.73}$$

A storage coil consisting of 2^n sections can now undergo a total of n parallel switching processes; in each step the inductance is divided into two equal parts which are then connected in parallel. Each step diminishes the inductance of the storage coil by a factor of four. By using the conservation of magnetic flux the current and the energy in the load coil after n parallel switching processes of the storage coil can be calculated to be

$$I_{2,n} = I_0 \frac{1}{1+x} \prod_{m=1}^{n} \left(1 + \frac{1}{1+x\,4^m}\right) \quad \text{with} \quad x := \frac{L_2}{L_1}, \tag{4.74}$$

$$w_{2,n} = \frac{W_{2,n}}{W_0} = \frac{x}{(1+x)^2} \prod_{m=1}^{n} \left(1 + \frac{1}{1+x\,4^m}\right)^2 . \tag{4.75}$$

Here I_0 and W_0 are the initial current and magnetic energy, respectively, in the storage coil, and x is the ratio of the load and storage inductance

Table 4.6. Resulting inductance for two inductances L_m and L_n with mutual inductance $M = k\sqrt{L_m L_n}$ connected in serial, L_{ser}, or parallel, L_{par}

	L_{ser}	L_{par}
general	$L_m + L_n + 2M$	$\dfrac{L_m L_n}{L_m + L_n} \dfrac{L_m + L_n + 2M}{L_m + L_n}$
$L_n = L_m = L$	$2(1+k)L$	$\dfrac{1+k}{2} L$
$k = 0$	$L_m + L_n$	$\dfrac{L_m L_n}{L_m + L_n}$
$k = 1$	$L_m + L_n + 2\sqrt{L_m L_n}$	$\dfrac{L_m L_n}{L_m + L_n} \dfrac{L_m + L_n + 2\sqrt{L_m L_n}}{L_m + L_n}$
$k = 0, L_m = L_n = L$	$2L$	$\dfrac{1}{2} L$
$k = 1, L_m = L_n = L$	$4L$	L

before the switching process. For practical reasons the number of switching steps is limited to two or even only one step, because each step requires three switches, which makes the circuit expensive. Additionally, the relative gain in efficiency is highest for the first step. In Fig. 4.18 the efficiencies for the one- and two-step switching scheme are drawn as a function of the inductance ratio $x = \frac{L_2}{L_1}$. For comparison the curve without parallel switching is also given. The latter has a maximum in relative efficiency of 0.25 at $x = 1$. The one-step mode has for all x values a higher efficiency with a maximum of 0.40 for an inductance ratio of $x = 0.5$. Still better results are achieved with the two-step mode, which gives for $x = 0.25$ an efficiency of 0.52. For the parallel switching scheme it is not necessary to have a mutual inductance between the sections of the coil, a feature which is essential in the serial switching scheme.

Serial Switching Scheme. Here the storage coil consists of numerous sections, which are successively switched out. One step of this serial switching scheme is shown in Fig. 4.19. Initially the switch S_1 is closed and S_2 is open, and a current I_0 flows. The storage coil consists of two sections $L_{1,\alpha}$ and $L_{1,\beta}$ with a mutual inductance M between them. Then S_2 is closed, the flux in the loop II is

$$\Phi_i = (L_{1,\beta} + M + L_2) I_0 . \tag{4.76}$$

Finally, after the switch S_1 is opened, the magnetic flux in the loop II is

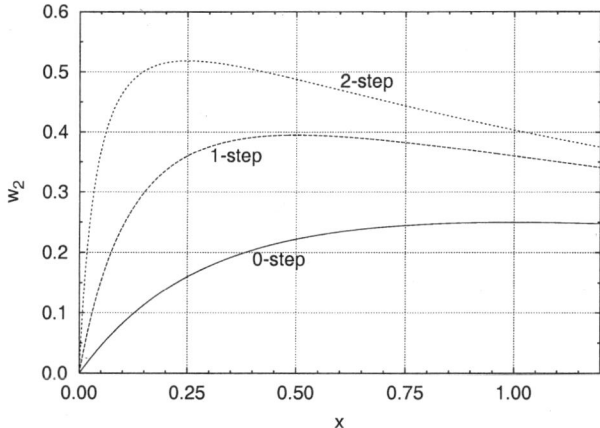

Fig. 4.18. Relative efficiency of the one-step and two-step mode of the parallel switching as a function of the initial inductance ratio $x = \frac{L_2}{L_1}$. For comparison the curve without parallel switching is also shown

$$\Phi_{\mathrm{f}} = (L_{1,\beta} + L_2) I_f \tag{4.77}$$

and because of flux conservation the final current is calculated to be

$$I_{\mathrm{f}} = I_0 \left(1 + \frac{M}{L_{1,\beta} + L_2}\right). \tag{4.78}$$

The switching process has increased the magnetic energy in the load L_2 from

$$W_0 = \frac{1}{2} L_2 I_0^2 \quad \text{to} \quad W_{\mathrm{f}} = W_0 \left(1 + \frac{M}{L_{1,\beta} + L_2}\right)^2. \tag{4.79}$$

The factor of increase, $\left(1 + \frac{M}{L_{1,\beta}+L_2}\right)^2$, is greater than one only if there exists a mutual inductance M between the sections of the storage coil. It can be shown that in the limit of many infinite small switching steps the efficiency approaches 100%. This multistep serial switching scheme is also known as a 'Meatgrinder circuit' [154–157], for a derivation of the rather lengthy efficiency calculations we refer to the literature detailed in the Reference Section.

4.3 Energy Sources

The following chapter gives an overview of several possible designs for energy supplies. The selection of a certain storage technique for a pulsed field facility has not only to meet the requirements of energy and power, but also has to obey financial and geographical constraints. The construction of power

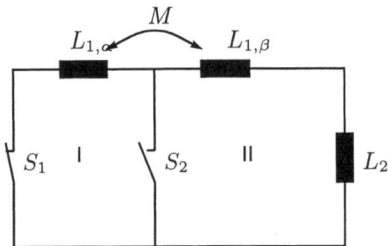

Fig. 4.19. One step of the serial switching scheme. The storage coil consists of two sections $L_{1,\alpha}$ and $L_{1,\beta}$ with a mutual inductance M. The switching process first closes the switch S_2 and then opens S_1. Because of the mutual inductance a part of the flux in the loop I is transferred into the loop II, which means an increase in magnetic energy there

supplies is a well-established branch of electrical engineering. This chapter will simply review the possible types without going too far into details and without being complete.

The power requirements depend strongly on the coil in use, and hence they show a wide variation. A first estimation for a high-field coil requires the energy source to store several megajoules of energy, releasing and transferring this energy with powers in the gigawatt range: A constant current density coil with inner bore diameter of 10 mm and coil parameters $\alpha = \beta = 20$ made from a hypothetical material with a tensile strength of 2.2 GPa could produce a central field of 100 T, requiring magnetic energy of 4.6 MJ. If we assume a rise time of the pulsed field of 5 ms, then the pulsed power source would need to deliver an average power of about 1 GW.

Possible types of energy sources are flywheel generators, batteries and capacitive and inductive storage systems; pioneering work was done by Kapitza [158–160]. Batteries are regarded as troublesome, because special types with low internal resistance are necessary. The same holds for generators, for each pulse requires high powers to be delivered, which implies rather overdimensioned machines. In that case a pulse causes only a relative small decrease of the flywheel's rotation speed. A full stop, though feasible, causes serious loads on the flywheel bearings and is therefore avoided under all circumstances. Capacitive storage systems are the most used today. Inductive storage systems are less frequently encountered, because the energy transfer can not as easily be realized as with capacitive methods [161–165]. A further possibility would be to place the pulsed field facility near to a power plant and use the power grid. Historically, pulsed magnets were pioneered by Peter Kapitza (1924). In his first experiments he used batteries as the energy source, and later he worked with generators. Inductive systems were also investigated, but found impracticable due to technical difficulties.

There is some work going on in the design of magnetic energy storage with superconducting coils. However, especially for pulsed power sources, these technique are not mature; see, for instance, [166–171] for a report on a small-scale system. As a source for a pulsed field facility superconducting magnetic energy storage technology does not seem suitable for now.

The design of the energy supply and the high-field coil have to be carried out simultaneously. The electrical parameters of the power strongly depend on the coil design, as an estimation with a capacitive storage system will show. The capacitor bank energizes an idealized coil with constant current density j_0 (see Sect. 4.1.1 with $R = 0$). If the current density j_0 is kept constant, then the coil generates the same central field and stores the same magnetic energy. Now the (average) current density can be realized either with few turns of wire with a large cross-section or many turns with a small cross-section. If we also want to have the same rise time of the field ($\omega_0 = 1/\sqrt{LC} = \text{const.}$), then the design with many turns of wire has to be paired with a rather low capacitance and vice versa. And since the energy of the circuit is transferred between the magnetic energy W_L of the coil and the electric energy W_C in the capacitor, this influences the load voltage of the latter:

$$\omega_0 = \frac{1}{\sqrt{LC}} = \text{const} \quad \text{and} \quad W_C = \frac{1}{2}CU^2 = \frac{1}{2}LI^2 = W_L . \qquad (4.80)$$

As a consequence a coil made with many turns of wire with small cross-section works with low currents and high voltages, and a solution with a thick wire and only a few turns requires large currents and low voltages.

Here we encounter some technical limitations: the voltage cannot be made deliberately high because of insulation problems, and the wire has to be bent to form the coil, which turns out to be difficult if the cross-section becomes too big.

Typical values encountered in laboratory installations work with voltages in the range 10 to 30 kV; for a power of 1 GW this would correspond to currents in the range 100 to 30 kA. For an overview of current high-field laboratories see [172–185].

This shows that the development of the coil and the energy source are interrelated. The optimal solution depends on the limitations of the coil (forces and heating) and the limitations of the power supply (voltage and current). The following sections deal with the various energy sources in more detail.

4.3.1 Local Power Grid

The direct supply from the local power grid is an economical solution if the high-field coil can be directly connected to the grid, i.e. the high-field facility lies close to a high-power line or to a power plant. As additional elements only a rectifier and eventually a transformer are necessary. The only technical limitation is the power one is allowed to draw from the power grid; energy is no issue in such a scenario.

Drawing power from the grid for only one half-wave of the line frequency causes severe distortions in the power supply, which the power companies are not very fond of. So they eventually set an upper limit for the power one is allowed to extract from the grid.

It depends very much on the local situation, which makes the power grid a viable option as the primary power supply; for instance, it was found to be an economical solution in Nijmegen [186]. At another site, the proposed high-field laboratory in Rossendorf near Dresden, the local grid turned out to be too limited in power [187].

4.3.2 AC Generator

Synchronous AC generators can operate successfully for many years in electric power plants. Typical voltages range from 1 kV to 15 kV. They are able to store great amounts of energy as kinetic energy in their rotating parts. The stored energy can be increased considerably by connecting a flywheel to the rotor.

Removing energy from of the generator means decelerating the rotation. A complete stop during one pulse is generally avoided because it would impose very strong forces on the bearings. This implies that the generator should store considerably more energy than that needed for one pulse [188–213].

Depending on the voltage required for the high-field coil a transformer may be necessary for adapting the generator voltage. Then the AC voltage has to be rectified. In order to reduce noise in the system, a 6-, 12- or even 24-pulse rectifying system with a corresponding number of thyristors will come to use. With a proper firing sequence of the thyristors and/or by regulating the excitation system of the generator some influence on the shape of the current pulse through the coil is possible. Typical prices for a complete system with generator, transformer and rectifier are [187]:

100 MW	300 MJ	$ 14 m
200 MW	600 MJ	$ 22 m
400 MW	1200 MJ	$ 31 m

The stored energies are fairly sufficient, but the power can be on the lower limit, especially when we look at the price! If a cheap generator is available and the power turns out to be too low, one can of course use the intermediate, inductive storage technique, as is done for instance at the High Magnetic Field Laboratory in Grenoble, France [214].

For the setup of a small pulsed field facility a generator is clearly not the best choice, because other storage methods are much cheaper. The advantages of the generator are the ability to change the pulse shape, for instance to increase the field in a stepwise manner. The disadvantages are the moderate power and the need for constant maintenance. The case may be different if one can rent a generator, as is done in Grenoble, or get a used one.

4.3.3 Battery

The storage of energy in chemical form in batteries has in recent years derived some benefits from new developments in lead-acid battery technology [215]. The inherent advantages of this new battery type are an increase in power and energy density, and by appropriate construction the inter-cell connection failure experienced in normal lead-acid batteries cannot occur. Since this battery type is mass produced for commercial applications, it is very cost effective.

Being based on electrochemical reactions the characteristic voltage is only a few volts, so that it is necessary to connect many cells in series to get voltages of at least a few 100 volts. Additionally, the internal resistance of these batteries is still quite high (several mΩ), so that to achieve a required output power a parallel connection becomes mandatory. In this way one easily ends up with an array of possibly thousands of batteries.

Installations with up to 700 V are known [215, 216], but this voltage level is still too low for a pulsed coil, which should preferably operate in the range above 10 kV. An intermediate inductor could resolve this issue, as demonstrated in [217]. The inductive storage coil is loaded at low voltages by the battery supply, and is then discharged into the high-field coil at a much higher voltage level.

The proposal [187] estimates the cost of a 35 MW battery supply working at 792 V as about $ 0.5 m. Again, an intermediate inductive storage coil is necessary in order to reach transfer voltages of several kV for transferring energy from the storage system into a high-field coil.

Though the low price seems very much in favor for an installation with batteries, there is still the problem with the opening switch at the inductive storage coil. Additionally one has to somehow incorporate safety features, because the total energy stored in the batteries is easily of the orders of several gigajoule. In the case of a short-circuit the current will practically flow for hours. Further safety concerns are the acid in the batteries, which requires that they are put into a special room so that acid from a damaged battery will not contaminate the environment. Maximising the lifetime of the batteries requires them to be operated in a very narrow temperature interval.

4.3.4 Homopolar Generator

The history of homopolar generators can be traced back to the early days of electrical research: Barlow studied copper disks rotating through the poles of magnets, and Faraday found a DC voltage generated between the rotor shaft and the periphery [218, 219]. The principle is shown in Fig. 4.20. More modern types of homopolar generator consist of a rotating drum rotating in a radial magnetic field. Because of its very simple and robust geometry the homopolar generator constitutes a very efficient energy storage system; typically it can store 10 to 100 MJ. The output power is very high because

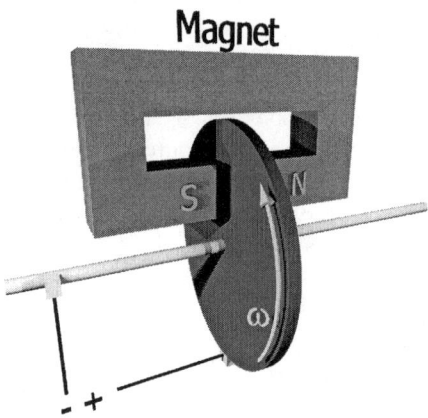

Fig. 4.20. Principle of a homopolar generator, also known as the Faraday disk or Barlow's wheel. A copper disk rotates between the poles of a magnet, generating a voltage between the axis and the periphery of the disk. The rather low voltages, typical values are 100 to 300 V, and output currents in the range of MA make the design of the brushes a technically difficult task

the decelerating electromagnetic forces are spread through the whole body of the rotor.

The voltage U generated by an element of length ℓ passing through a constant magnetic field B with constant velocity v is (B, v and ℓ are assumed orthogonal to each other)

$$U = v B \ell . \tag{4.81}$$

For a velocity of $v = 100 \, \text{m s}^{-1}$, a field of $B = 1 \, \text{T}$ and a length of $\ell = 1 \, \text{m}$ we get a voltage of $U = 100 \, \text{V}$.

The main disadvantages are the rather low output voltages, typically only 100–300 V DC, and very high output currents, easily in the MA range. Because of the low output voltages intermediate inductive energy storage is mandatory when using a homopolar generator for a pulsed coil, and once again one has to operate with an opening switch under an inductive load. The low voltages and the high currents also make the design of the current collection system a technically difficult task.

An historical overview of homopolar generators can be found in [220]. For descriptions of some installations, i.e. homopolar generators together with intermediate inductive storage and an opening switch, we refer to [221–233].

Looking only at stored energy and power, the homopolar generator looks very promising, the low voltage and the high output current are a severe drawback, however.

4.3.5 Inductive Energy Storage

Pulsed inductors have been used for intermediate energy storage and as a pulse compression component. Any low-voltage source can be used to drive current into an inductor, for instance batteries, homopolar generators or the local power grid. Getting the energy into the inductor is relatively easy, but getting the energy from the intermediate storage coil into the high-field coils requires an opening switch suitable for opening an inductive load (see Sect. 4.2).

An overview of ready-to-use systems, i.e. a storage coil with switching equipment, can be found in [161] and [234]. A comparison of large-scale SMES systems (Superconducting Magnetic Energy Storage) is given in [235]; for a few descriptions of storage coil techniques we refer to [166, 236–246] and for switch technologies to [247–269].

4.3.6 Capacitor Bank

Capacitive energy storage is often the preferred method for laboratory research. Capacitors are versatile and easy to use and very well matched to inductive loads. A typical storage systems consists of a large number of individual capacitors, closing switches, usually thyristors, and protection inductors, which limit the short-circuit currents. The deliverable power is limited mainly by the maximum current of the thyristors.

Table 4.7. Specification of the pulse capacitor type 32776 from General Atomics Energy Products (formerly Maxwell Energy Products, Inc.) [285]

Parameter	Value
Capacitance	$1667\,\mu\text{F}$
Tolerance	$\pm 10\,\%$
Rated voltage	$10.0\,\text{kV}$
Test voltage	$11.0\,\text{kV}$
Rated energy	$83.3\,\text{kJ}$
Rated voltage reversal	$30\,\%$
Max. voltage reversal	$80\,\%$
Rated peak current	$60\,\text{kA}$
Max. peak current	$120\,\text{kA}$
Rated design life	$10\,000$ cycles
DC Life	42 hours
Reliability at rated life	$90\,\%$
Dimensions	$343 \times 406 \times 787\,\text{mm}^3$
Weight	$154\,\text{kg}$

The data for a modern pulse capacitor are shown in Table 4.7, and more extensive descriptions and applications can be found in [270–284]. If ordered in sufficient quantities a capacitor bank can be as cheap as $ 70 k per MJ stored energy, [187].

The capacitors in Table 4.7 are a specially designed type for millisecond discharge capacitor banks. At a voltage of 10 kV one single capacitor can store an energy of 83 kJ, and typical energy densities are in 0.75 MJ m^{-3} range. They are self-healing metallized-electrode capacitors, which means that in the course of normal use they will gradually lose capacity. The design life was defined as the number of charge/discharge cycles at which a capacitance loss of 5% occurs. For the type 32776, a load voltage of 10 kV, a voltage reversal of 30% and a discharge peak current of 25 kA this can be expected to happen after 10 000 charge/discharge cycles. At that point the capacitor is not damaged, but has only lost 5% of its original capacitance. A dangerous rupture of the capacitor can be expected at a capacitance loss of about 10%. The lifetime increases considerably when working at lower load voltages. At 9 kV, for instance, the number of cycles reaches 60 000, and at 8 kV more than 350 000 cycles are possible.

4.3.7 Conclusion

The energy source for a pulsed coil has to meet certain requirements regarding transferable energy (several MJ) and power (GW). The four possible principles are capacitive, inductive, chemical and kinetic energy storage. The achievable energy density for present state-of-the-art data are given in

Table 4.8. Comparison of possible energy storage techniques. For each principle the expression for the energy density is given, which results with some assumptions and present state-of-the-art data in a corresponding value for the energy density. (Table adapted from [188]; for data for the Pulsar battery see [216])

Device	Capacitor	Rotor	Inductor	Battery
Method	Electrostatic	Inertial	Magnetic	Chemical
Equation	$W = \frac{1}{2}\varepsilon E^2$	$W = \frac{1}{2}I_m\omega^2$	$W = \frac{1}{2\mu_0}B^2$	
Assumption	High energy density plastic film	High-speed rotor	High-field air-cored inductor	Type Pulsar^{TM}
State of the art	$E = 400$ V m^{-1}, $\varepsilon_r = 10$	$\varrho = 1500$ kg m^{-3}, $v = 600$ m s^{-1}	$B = 40$ T	
Energy density [MJ m^{-3}]	7	135	640	1300

Table 4.9. Comparison of various energy storage methods (see also [286])

Method	Power grid	AC generator	Battery	Homopolar generator	Capacitor bank
Investment	modest	high	low	modest	modest
Noise	high	modest	small	modest	very small
Feedback to grid	high	modest	none	small	none
Running costs	low	very high	high	very high	low
Energy	infinite	high	infinite	high	high
Power	limited	limited	limited	high	very high
Remark			inductor necessary	inductor necessary	

Table 4.8. The highest energy density is reached with the electrochemical method in batteries, the lowest one with capacitors. As seen before, batteries are not well suited for pulsed coils, however, and intermediate inductive storage is necessary due to its intrinsic high internal resistance.

Selecting an optimal energy storage system for a pulsed magnetic field facility must therefore consider not only the storage technique, but also has to take into account pulse-forming networks, switches, maintenance costs, versatility and also geographical issues.

The rather low voltage of the storage techniques using the battery and the homopolar generator make intermediate storage in an inductor mandatory, which leads to the problems of an opening switch under an inductive load, as outlined in Sect. 4.2. Of course, inductive intermediate storage can also be used in conjunction with the power grid or an AC generator, if one needs higher powers. The inductive energy storage system works in such a case as a power converter.

We close with Table 4.9, which compares the different storage mechanisms with respect to the initial investment, the electrical noise impact on the high-field coil and therefore the experimental conditions, and the noise feedback to the local power grid. Furthermore, we give indications of the running costs and the energy and power one can expect.

References

1. K. Kitazawa, Y. Ikezoe, H. Uetake, N. Hirota: Physica B **294-295**, 709 (2001)
2. E. Beaugnon, F. Fabregue, D. Billy, J. Nappa, R. Tournier: Physica B **294-295**, 715 (2001)
3. J.S. Brooks, J.A. Cothern: Physica B **294-295**, 721 (2001)
4. M. Motokawa, M. Hamai, T. Sato, I. Mogi, S. Awaji, K. Watanabe, N. Kitamura, M. Makihara: Physica B **294-295**, 729 (2001)
5. A.K. Geim, H.A.M.S. ter Tisha: Physica B **294-295**, 736 (2001)
6. J.D. Jackson: *Classical Electrodynamics*, 2nd edn. (John Wiley, New York 1962)
7. W.R. Smythe: *Static and Dynamic Electricity*, 3rd edn. (McGraw-Hill, New York 1968)
8. J.J. Roche: Am. J. Phys. **68**, 438 (2000)
9. J. Crangle, M. Gibbs: Physics World, **7**(11), 31 (1994)
10. F.W. Grover: *Inductance Calculations: Working Formulas and Tables* (Dover Publications, New York 1946)
11. S. Butterworth: Philos. Mag. **29**, 578 (1915)
12. S. Butterworth: Philos. Mag. **31**, 443 (1916)
13. A. Campbell: Philos. Mag. **15**, 155 (1908)
14. H.B. Dwight: *Electrical Coils and Conductors* (McGraw-Hill, New York 1945)
15. J.V. Jones: Philos. Mag. **27**, 56 (1889)
16. K. Kim, E. Levi, Z. Zabar, L. Birenbaum: IEEE Trans. Magn. **32**, 478 (1996)
17. T.R. Lyle: Philos. Trans. **213A**, 421 (1913)
18. H.A. Wheeler: Proc. I.R.E. 412 (1942)
19. J.C. Maxwell: *Electricity and Magnetism* (Dover Publications)
20. A. Rezzoug, J. P. Caron, F. M. Sargos: IEEE Trans. Magn. **28**, 2250 (1992)
21. T.L. Simpson: IEEE Trans. Magn. **35**, 508 (1999)
22. W. Mai, G. Henneberger: IEEE Trans. Magn. **35**, 1590 (1999)
23. S. Babic, C. Akyel: IEEE Trans. Magn. **36**, (2000), 1970
24. M. de Almeida Bueno, A.K.T. Assis: IEEE Trans. Magn. **34**, 317 (1998)
25. M.W. Garrett: J. Appl. Phys. **22**, 1091 (1951)
26. M.W. Garrett: J. Appl. Phys. **34**, 2567 (1963)
27. M.W. Garrett: J. Appl. Phys. **38**, 2563 (1967)
28. M.W. Garrett: J. Appl. Phys. **40**, 3171 (1969)
29. A.I. Rusinov: IEEE Trans. Magn. **30**, 2685 (1994)
30. R. Reeves: J. Phys. E, Sci. Instrum. **21**, 31 (1988)
31. N.J. Diserens: IEEE Trans. Magn. **19**, 2304 (1983)
32. B. Azzerboni, E. Cardelli, A. Tellini: IEEE Trans. Magn. **25**, 4462 (1989)
33. I.R. Ciric: IEEE Trans. Magn. **24**, 3132 (1988)
34. I.R. Ciric: IEEE Trans. Magn. **27**, 669 (1991)
35. I.R. Ciric: IEEE Trans. Magn. **28**, 1064 (1992)
36. C. Hafner, R. Ballisti: IEEE Trans. Magn. **25**, 2828 (1989)

37. T. Onuki, S. Wakao: IEEE Trans. Magn. **31**, 1476 (1995)
38. C.F. Weggel, D.P.: IEEE Trans. Magn. **24**, 1544 (1988)
39. J.C.-E. Sten: IEEE Trans. Magn. **34**, 199 (1998)
40. A. Formisano, R. Martone, F. Villone: IEEE Trans. Magn. **34**, 218 (1998)
41. C. Yuesen: IEEE Trans. Magn. **34**, 502 (1998)
42. A. Nicolaide: IEEE Trans. Magn. **34**, 608 (1998)
43. K. Ozaki, M. Kobayashi, G. Rowlands: IEEE Trans. Magn. **34**, 2185 (1998)
44. B. Azzerboni, G.A. Saraceno, E. Cardelli: IEEE Trans. Magn. **34**, 2601 (1998)
45. M. Feliziani, F. Maradei: IEEE Trans. Magn. **34**, 2795 (1998)
46. S. Gyimothy, I. Sebestyen: IEEE Trans. Magn. **34**, 3427 (1998)
47. J. Kangas, T. Tarhasaari, L. Kettunen: IEEE Trans. Magn. **36**, 1645 (2000)
48. P. Konzbul, K. Sveda, A. Srnka: IEEE Trans. Magn. **36**, 1732 (2000)
49. I.R. Ciric: IEEE Trans. Magn. **36**, 1990 (2000)
50. L.K. Urankar: IEEE Trans. Magn. **18**, 1860 (1982)
51. L.K. Urankar: IEEE Trans. Magn. **26**, 1171 (1990)
52. L.K. Urankar, P. Henninger, F.S. Nestel: IEEE Trans. Magn. **30**, 1236 (1994)
53. M. Abramowitz, I.A. Stegun, *Handbook of Mathematical Functions* (Dover Publications, New York 1972)
54. W. Bartky: Rev. Mod. Phys. **10**, 264 (1938)
55. A. Sommerfeld: *'Mechanik der deformierbaren Medien'*, Nachdruck der 6. Auflage (Verlag Harri Deutsch, Thun und Frankfurt/Main 1978)
56. M. Iremonger: *Basic Stress Analysis* (Butterworth Scientific, London 1982)
57. R. Roark: *Formulas for Stress and Strain* (McGraw-Hill, New York 1965)
58. S.H. Crandall, N. Dahl, T. Lardner: *An Introduction to the Mechanics of Solids* (McGraw-Hill, New York 1978)
59. E. Hearn: *Mechanics of Materials* (Pergamon Press, New York 1977)
60. A.E.H. Love: *A Treatise on the Mathematical Theory of Elasticity* (Dover, New York 1944)
61. M.N. Wilson, *Superconducting magnets* (Clarendon Press, Oxford 1983)
62. P.O. Carden: J. Phys. E, Sci. Instrum. **5**, 654 (1972)
63. P.O. Carden: J. Phys. E, Sci. Instrum. **5**, 657 (1972)
64. P.O. Carden: J. Phys. E, Sci. Instrum. **5**, 663 (1972)
65. P.O. Carden: J. Phys. E, Sci. Instrum. **5**, 667 (1972)
66. P.O. Carden: J. Phys. E, Sci. Instrum. **7**, 750 (1974)
67. A.W. Cox, H. Garmestani, W.D. Markiewicz, I.R. Dixon: IEEE Trans. Magn. **32**, 3012 (1996)
68. H. Garmestani, M.R. Vaghar, W.D. Markiewicz: IEEE Trans. Magn. **30**, 2237 (1994)
69. J.L. Hill, B.C. Amm, J. Schwartz: IEEE Trans. Magn. **30**, 2094 (1994)
70. H.P. Furth, S.C. Jardin, D.B. Montgomery: IEEE Trans. Magn. **24**, 1467 (1988)
71. W.D. Markiewicz, M.R. Vaghar, I.R. Dixon, H. Garmestani: IEEE Trans. Magn. **30**, 2233 (1994)
72. Y. Miura, M. Sakota, R. Shimada: IEEE Trans. Magn. **30**, 2573 (1994)
73. G.A. Shneerson: IEEE Trans. Magn. **28**, 505 (1992)
74. L. Kettunen, K. Forsman, A. Bossavit: IEEE Trans. Magn. **34**, 2551 (1998)
75. R. Hill, *The Mathematical Theory of Plasticity* (Clarendon Press, Oxford 1971)
76. L.M. Kachanov: *Foundations of the Theory of Plasticity* (North-Holland, Amsterdam 1971)
77. J. Lemaitre, J.-L. Chaboche: *Mechanics of Solid Materials* (Cambridge University Press, Cambridge 1990)

References

78. S. Timoshenko: *Strength of Materials*, 3rd edn. (van Nostrand, New York 1958)
79. S. Timoshenko: *History of Strength of Materials* (McGraw-Hill, New York 1953)
80. N.W. Ashcroft, N.D. Mermin: *Solid State Physics* (Holt Saunders Japan, Tokyo 1981)
81. N.J. Simon, E.S. Drexler, R.P. Reed: *Properties of Copper and Copper Alloys at Cryogenic Temperatures* (NIST Monograph **177**, Boulder 1992)
82. D.B. Montgomery: *Solenoid Magnet Design* (Wiley-Interscience, New York 1969)
83. D.B. Montgomery: Rep. Prog. Phys. **26**, 69 (1963)
84. F. Herlach: Rep. Prog. Phys. **62**, 859 (1999)
85. D. De Klerk: *The Construction of High-Field Electromagnets* (Newport Instruments Ltd., Newport Pagnell 1965)
86. D.J. Kroon: *Laboratory Magnets* (Philips Gloeilampenfabriken, Eindhoven, The Netherlands 1968)
87. D.B. Montgomery: J. Appl. Phys. **36**, 893 (1965)
88. J.D. Cockroft: Philos. Trans. R. Soc. London A **227**, 317 (1928)
89. F. Bitter: Rev. Sci. Instrum. **7**, 482 (1936)
90. M. Wood, 'The High-Field Magnets At Oxford'. In: *High Magnetic Fields* (MIT Press, Cambridge Mass., 1961) pp. 387-392
91. H.-J. Schneider-Muntau: IEEE Trans. Magn. **24**, 1041 (1988)
92. H.-J. Schneider-Muntau, S. Prestemon: IEEE Trans. Magn. **28**, 486 (1992)
93. H.-J. Schneider-Muntau, J.C. Vallier: IEEE Trans. Magn. **24**, 1067 (1988)
94. M. Date: IEEE Trans. Magn. **12**, 1024 (1976)
95. J.O. Ratka, W.D. Spiegelberg: IEEE Trans. Magn. **30**, 1859 (1994)
96. W. Heller: *Copper-Beryllium Alloys for Technical Applications*, (CERN, Geneva 1976)
97. AerMet 100, Alloy developed by Carpenter Technology Corp., Reading, PA, USA
98. R.J. Weggel, J.O. Ratka, W.D. Spiegelberg, Y. Sakai: IEEE Trans. Magn. **30**, 2188 (1994)
99. B.P. Bardes (Ed.): *Metals Handbook, 9th edn. vol.1* (American Society for Metals, Metals Park, Ohio 1978)
100. R.F. Decker: *Source Book on Maragin Steels* (American Society for Metals, Metals Park, Ohio 1979)
101. www-unix.mcs.anl.gov/otc/Guide/faq/nonlinear-programming-faq.html
102. C.R.I. Emson, J. Simkin, C.W. Trowbridge: IEEE Trans. Magn. **30**, 1533 (1994)
103. csep1.phy.ornl.gov/mo/mo.html
104. M. Chiampi, C. Ragusa, M. Repetto: IEEE Trans. Magn. **32**, 1234 (1996)
105. E. Coccorese, R. Martone, F.C. Morabito: IEEE Trans. Magn. **30**, 2829 (1994)
106. J.M. Dohlus, P. Hahne, X. Du, B. Wagner, T. Weiland, S.G. Wipf: IEEE Trans. Magn. **29**, 1914 (1993)
107. A. Gottvald, K. Preis, C. Magele, O. Biro, A. Savini: IEEE Trans. Magn. **28**, 1537 (1992)
108. J.-D. Kant, J. Le Drezen, J. Bigeon: IEEE Trans. Magn. **31**, 1968 (1995)
109. R.J. Lari: IEEE Trans. Magn. **18**, 316 (1982)
110. J. Simkin, C.W. Trowbridge: IEEE Trans. Magn. **28**, 1545 (1992)
111. M. Kitamura, S. Kakukawa: IEEE Trans. Magn. **34**, 2908 (1998)
112. F. Wurtz, J. Bigeon, J.L. Coulomb, C. Espanel, J.M. Kauffmann: IEEE Trans. Magn. **34**, 3411 (1998)
113. http://bedvgm.kfunigraz.ac.at:8001/alex/solvopt/

114. Z.J. Cendes: IEEE Trans. Magn. **27**, 3958 (1991)
115. Z.J. Cendes, D. Shenton, H. Shahnasser: IEEE Trans. Magn. **19**, 2551 (1983)
116. I.E. Lager, G. Mur: IEEE Trans. Magn. **32**, 631 (1996)
117. E.J. Silva, R.C. Mesquita, R.R. Saldanha,P.F.M. Palmeira: IEEE Trans. Magn. **30**, 3618 (1994)
118. M.V.K. Chari, P.P. Silvester: *Finite Elements in Electrical and Magnetic Field Problems* (John Wiley, New York 1980)
119. PDEase, Macsyma Inc., 20 Academy Street, Arlington, MA 02174, U.S.A.
120. K. Budinski: *Engineering Materials Properties and Selection*, 3d edn. (Prentice Hall, New Jersey 1989)
121. D. Chung: *Carbon Fiber Composites* (Butterworth-Heinemann, Boston 1994)
122. M. Schwartz: *Composite Materials Handbook* (McGraw-Hill, New York 1984)
123. E.L. Danfelt, S. A. Hewes, T. W. Chou: Int. J. Mech. Sci. **19**, 69 (1977)
124. G.A. Cuccuru, F. Ginesu, B. Picasso, P. Priolo: J. Compo. Mater. **14**, 31 (1980)
125. www.toyobo.co.jp/e/seihin/kc/pbo/
126. private communication: B. Lesch, NHMFL Tallahassee
127. K. Rosseel, F. Herlach, W. Boon, Y. Bruynseraaede: Physica B **294-295**, 657 (2001)
128. W.H. Press, B.P. Flannery, S.A. Teukolsky, W.T. Vetterling: *Numerical Recipes in Pascal* (Cambridge Univ. Press, Cambridge 1989)
129. X.-W. Dai, R. Ludwig, R. Palanisamy: IEEE Trans. Magn. **26**, 3089 (1990)
130. K.R. Davey, L. Turner: IEEE Trans. Magn. **26**, 1164 (1990)
131. L.R. Egan, E.P. Furlani: IEEE Trans. Magn. **27**, 4343 (1991)
132. Z.Z. Feng: IEEE Trans. Magn. **21**, 993 (1985)
133. S.M. Haugland: IEEE Trans. Magn. **32**, 3195 (1996)
134. A. Konrad: IEEE Trans. Magn. **21**, 1805 (1985)
135. A.B. Kos, F.R. Fickett: IEEE Trans. Magn. **30**, 4560 (1994)
136. J.S. Lee: IEEE Trans. Magn. **27**, 4000 (1991)
137. M.E. Lee, S.I. Hariharan, N. Ida: J. Comput. Phys. **89**, 319 (1990)
138. L. Li, F. Herlach: J. Phys. D, Appl. Phys. **31**, 1320 (1998)
139. I.V. Lindell, E.A. Lehtola, K.I. Nikoskinen: IEEE Trans. Magn. **29**, 2202 (1993)
140. R. Ludwig, X.-W. Dai: IEEE Trans. Magn. **26**, 299 (1990)
141. A. Nishikata, S. Kiener: IEEE Trans. Magn. **30**, 3359 (1994)
142. E.J.W. ter Maten, J.B.M. Melissen: IEEE Trans. Magn. **28**, 1287 (1992)
143. H. Tsuboi, M. Tanaka, T. Misaki: IEEE Trans. Magn. **26**, 1647 (1990)
144. M. Tsuchimoto, K. Miya, A. Yamashita, M. Hashimoto: IEEE Trans. Magn. **28**, 1434 (1992)
145. K. Halbach: Nucl. Instrum. Methods **107**, 529 (1973)
146. E. Fraga, C. Prados, D.-X. Chen: IEEE Trans. Magn. **34**, 205 (1998)
147. E. Uzal, I. Ozkol, M.O. Kaya: IEEE Trans. Magn. **34**, 213 (1998)
148. I.D. Mayergoyz: IEEE Trans. Magn. **34**, 1228 (1998)
149. E.J. Escorcia-Aparicio, H.J. Choi, W.I. Ling, R.K. Kawakami, Z.Q. Qiu, H. Tsuboi, K. Ikeda, M. Kurata, K. Kainuma, K. Nakamura: IEEE Trans. Magn. **34**, 1234 (1998)
150. P.J. Wang, S.J. Chiueh: IEEE Trans. Magn. **34**, 1237 (1998)
151. P.C. Rem, E.M. Beunder, A.J. van den Akker: IEEE Trans. Magn. **34**, 2280 (1998)
152. M. Uesaka, K. Hakuta, K. Miya, K. Aoki, A. Takahashi: IEEE Trans. Magn. **34**, 2287 (1998)
153. M. Hecquet, P. Brochet, S.J. Lee, P. Delsalle: IEEE Trans. Magn. **35**, 1841 (1999)

154. D. Giorgi, J. Long, T. Navapanich, K. Lindner, A. Griffin, O. Zucker: IEEE Trans. Magn. **22**, 1485 (1986)
155. K. Lindner, J. Long, T. Navapanich, O. Zucker: IEEE Trans. Magn. **22**, 1591 (1986)
156. J. Long, K. Lindner, O. Zucker: 'Analysis and Comparison of Circuits Undergoing a Change of Inductance via Continuous Sequential Switching and/or Geometrical Change'. In: *Megagauss Technology and Pulsed Power Applications* (Plenum Press, New York 1987) pp. 596-607
157. O. Zucker, J. Wyatt, K. Lindner: IEEE Trans. Magn. **20**, 391 (1984)
158. P. Kapitza: Proc. Roy. Soc. Lond. A **105**, 691 (1924)
159. P. Kapitza: Proc. Roy. Soc. Lond. A **115**, 658 (1927)
160. P. Kapitza: Usp. Fiz. Nauk **163**, 77 (1993)
161. A.S. Druzhimin, V.G. Kuchinsky, B.A. Larionov, A.G. Roshal, V.P. Silin, V.F. Soikin: IEEE Trans. Magn. **28**, 410(1992)
162. Ch. Maissonnier, J.G. Linhart, C. Gourlan: Rev. Sci. Instrum. **37**, 1380 (1966)
163. O. Zucker, J. Long, K. Lindner, D. Giorgi, T. Navapanich: Rev. Sci. Instrum. **57**, 859 (1986)
164. A.J. Moses: IEEE Trans. Magn. **34**, 1186 (1998)
165. E. Steingroever, G. Ross: IEEE Trans. Magn. **34**, 2084 (1998)
166. W. Weck, P. Ehrhart, A. Muller, G. Reiner: IEEE Trans. Magn. **33**, 524 (1997)
167. K.V. Dubovenko, V.T. Chemerys: IEEE Trans. Magn. **35**, 328 (1999)
168. W. Weck, P. Ehrhart, A. Muller, H. Scholderle, E. Sturm: IEEE Trans. Magn. **35**, 383 (1999)
169. K. Ryu, H.J. Kim, K.C. Seong, J.W. Cho, E.Y. Lee, H.B. Jin, K.S. Ryu: IEEE Trans. Magn. **35**, 4103 (1999)
170. C.A. Borghi, M. Fabbri, P.L. Ribani: IEEE Trans. Magn. **35**, 4275 (1999)
171. J. Biebach, P. Ehrhart, A. Muller, G. Reiner, W. Weck: IEEE Trans. Magn. **37**, 353 (2001)
172. F. Herlach: Physica B **294-295**, 500 (2001)
173. B.L. Brandt, S. Hannahs, H.-J. Schneider-Muntau, G. Boebinger, N.S. Sullivan: Physica B **294-295**, 505 (2001)
174. G.S. Boebinger, H.H. Lacerda, H.-J. Schneider-Muntau, N. Sullivan: Physica B **294-295**, 512 (2001)
175. F. Debray, H. Jongbloets, W. Joss, G. Martinez, E. Mossang, J.C. Picoche, A. Plante, P. Rub, P. Sala, P. Wyder: Physica B **294-295**, 523 (2001)
176. J. Perenboom, R. Peters, T. Roeterdink, S. Wiegers, P. Zwinkels, P. Frings, J.-K. Maan: Physica B **294-295**, 529 (2001)
177. T. Kiyoshi, S. Matsumoto, T. Asano, H. Wada: Physica B **294-295**, 535 (2001)
178. K. Watanabe, S. Awaji, K. Takahashi, M. Motokawa: Physica B **294-295**, 541 (2001)
179. O. Portugall, F. Lecoutourier, J. Marquez, D. Givord, S. Askenazy: Physica B **294-295**, 579 (2001)
180. K. Kindo: Physica B **294-295**, 585 (2001)
181. J. Vanacken: Physica B **294-295**, 591 (2001)
182. H. Jones, W.J. Siertsema, P.E. Richers, M. Newson, P.M. Saleh, A.L. Hickman: Physica B **294-295**, 598 (2001)
183. H. Krug, M. Doerr, D. Eckert, H. Eschrig, F. Fischer, P. Fulde, R. Groessinger, A. Handstein, F. Herlach, D. Hinz, R. Kratz, M. Loewenhaupt, K.H. Müller, F. Pobell, L. Schultz, H. Siegel, F. Steglich, P. Verges: Physica B **294-295**, 605 (2001)
184. *Proc. European Workshop Science in 100 T*, Leuven, May 15-17, 1992, ed. by L.J. Challis, J.J.M. Franse, F. Herlach, P. Wyder (K. U. Leuven, Belgium, 1992)

185. *Proc. EUPRO Workshop Science in 100 T II*, Leuven, September 31 - October 1, 1994, ed. by L.J. Challis, J.J.M. Franse, F. Herlach, P. Wyder (K. U. Leuven, Belgium 1994)
186. J.A.A.J. Perenboom, J.C. Maan, S.A.J. Wiegers, P.H. Frings: IEEE Trans. Appl. Supercond. **10**, 1549 (2000)
187. Vorschlag zur Errichtung eines Labors für gepulste, sehr hohe Magnetfelder in Dresden (Hochfeldlabor Dresden, 1999); Web: www.fz-rossendorf.de/HLD/
188. I.R. McNab: IEEE Trans. Magn. **33**, 453 (1997)
189. L.A. Kilgore, E.J. Hill, C. Flick: IEEE Trans. Power Appar. Syst. **82**, 442 (1963)
190. D.J. Scott, R.M. Calfo: Synchronous Machines for Pulsed Power Applications. In: *7th IEEE Pulsed Power Conf., Monterey, California, 1989*, ed. by B.H. Bernstein, J.P. Shannon (IEEE, New York 1989) pp. 229-232
191. R.L. Puterbaugh, P.P. Mongeau, B. Acharya, D. Curtiss: IEEE Trans. Magn. **31**, 73 (1995)
192. J.R. Kitzmiller, R.W. Faidley, R.L. Fuller, R.N. Hedifen, R.F. Thelan: IEEE Trans. Magn. **29**, 441 (1993)
193. W.A. Walls, S.B. Pratap, W.G. Brinkman, K.G. Cook, J.D. Herbst, S.M. Manifold, B.M. Rech, R.F. Thelan, R.C. Thompson: IEEE Trans. Magn. **27**, 335 (1991)
194. R.M. Calfo, D.J. Scott, D.W. Scherbarth, M.T. Buttram, H.C. Harjes: Design and Test of a Continuous Duty Pulsed AC Generator. In: *8th IEEE Int. Pulsed Power Conf., San Diego, California, 1991*, ed. by R. White, K. Prestwich (IEEE, New York 1991) pp. 715-718
195. D.J. Scott, R.M. Calfo, H.R. Schwenk: Development of High Power Density Pulsed AC Generator. In: *8th IEEE Int. Pulsed Power Conf., San Diego, California, 1991*, ed. by R. White, K. Prestwich (IEEE, New York 1991) pp. 549-552
196. B.T. Murphy, S.M. Manifold, J.R. Kitzmiller: IEEE Trans. Magn. **33**, 474 (1997)
197. S.B. Pratap, J.P. Kajs, W.A. Walls, W.F. Weldon, J.R. Kitzmiller: IEEE Trans. Magn. **33**, 495 (1997)
198. G. Genta: *Kinetic Energy Storage Theory and Practice of Advanced Flywheel Systems* (Butterworths, London 1985)
199. O. Mahrenholtz (Ed.): *Dynamics of Rotors: Stability and System Identification* (Springer-Verlag, Wien 1984)
200. J.C. Georgian: J. Compos. Mater. **23**, 2 (1989)
201. R.D. Blevins: *Formulas for Natural Frequency and Mode Shapes* (R.E. Krieger, Malabar Florida 1984)
202. D. Childs: *Turbomachinery Rotordynamics* (John Wiley, New York 1993)
203. F. Ehrich (Ed.): *Handbook of Rotordynamics* (McGraw-Hill, New York 1992)
204. K.G. Cook, B.T. Murphy, S.M. Manifold, T. Pak, M.D. Werst, J.R. Kitzmiller, W.A. Walls, A. Alexander, K. Twigg: IEEE Trans. Magn. **35**, 277 (1999)
205. G. Nagy, S. Rosenwasser, G. Mehle: IEEE Trans. Magn. **35**, 289 (1999)
206. S. Rosenwasser, G. Nagy, G. Mehle: IEEE Trans. Magn. **35**, 307 (1999)
207. R. Acebal: IEEE Trans. Magn. **35**, 317 (1999)
208. T.A. Aanstoos, J.P. Kajs, W.G. Brinkman, H.P. Liu, A. Ouroua, R.J. Hayes, C. Hearn, J. Sarjeant, H. Gill: IEEE Trans. Magn. **37**, 242 (2001)
209. A. Balikci, Z. Zabar, D. Czarkowski, E. Levi, L. Birenbaum: IEEE Trans. Magn. **37**, 280 (2001)
210. J.H. Beno, R.C. Thompson, M.D. Werst, S.M. Manifold, J.J. Zierer: IEEE Trans. Magn. **37**, 284 (2001)

References 175

211. M.D. Driga, S.B. Pratap, A.W. Walls, J.R. Kitzmiller: IEEE Trans. Magn. **37**, 295 (2001)
212. J.T. Tzeng: IEEE Trans. Magn. **37**, 328 (2001)
213. I.R. McNab: IEEE Trans. Magn. **37**, 375 (2001)
214. Grenoble High Magnetic Field Laboratory, 25 Avenue des Martyrs, F-38042 Grenoble Cedex 9, France
215. J.R. Lippert: IEEE Trans. Magn. **29**, 1009 (1993)
216. J.P. Kajs, R.C. Zowarka: IEEE Trans. Magn. **29**, 1003 (1993)
217. J.A. Pappas, G.R. Headifen, J.M. Weldon, J.C. Wright, R.C. Zowarka, T.A. Aanstoos, J.H. Kajs: IEEE Trans. Magn. **29**, 1027 (1993)
218. G. Forbes: 'Dynamo-electric machine', US Patent No. 338 169 (1886)
219. N. Tesla: 'Notes on a Unipolar Dynamo'. In: *The Electrical Engineer* (New York, Sept. 2, 1891), Reprinted in: *The Inventions, Researches and Writings of Nikola Tesla* (Barnes and Noble, New York 1995)
220. I.R. McNab: IEEE Trans. Magn. **33**, 461 (1997)
221. I.M. Vitkovitsky, R.D. Ford, D. Jenkins, W.H. Lupton: IEEE Trans. Magn. **18**, 157 (1982)
222. G.R. Headifen, J.A. Pappas, J.M. Weldon, J.C. Wright, J.H. Price, J.H. Gully, G. Brunson: IEEE Trans. Magn. **29**, 980 (1993)
223. D.H. Curtiss, P.P. Mongeau, R.L. Puterbaugh: IEEE Trans. Magn. **31**, 26 (1995)
224. M.D. Driga, S.B. Pratap, W.F. Weldon: IEEE Trans. Magn. **25**, 142 (1989)
225. R.L. Fuller, J.R. Kitzmiller, R.F. Thelen: IEEE Trans. Magn. **31**, 52 (1995)
226. A.J. Mitcham, D.H. Prothero, J.C. Brooks: IEEE Trans. Magn. **25**, 362 (1989)
227. S.B. Pratap, K.T. Hsieh, M.D. Driga, W.F. Weldon: IEEE Trans. Magn. **25**, 454 (1989)
228. N.D. Baker, B.D. McKee, I.R. McNab: IEEE Trans. Magn. **22**, 1386 (1986)
229. T.M. Bullion, M. D. Driga, J. H. Gully, H. G. Rylander, K. M. Tolk, W. F. Weldon, H. H. Woodson, R. Zowarka: Testing and Analysis of a Fast Discharge Homopolar Machine (FDX)'. In: *2nd IEEE Int. Pulse Power Conf., Lubbock, Texas 1979*, ed. by A. H. Guenther, M. Kristiansen (IEEE, New York 1979) pp. 333-342
230. W.A. Walls, Wm.F. Weldon, M.D. Driga, S.M. Manifold, H.H. Woodson, J.H. Gully: IEEE Trans. Magn. **22**, 1793 (1986)
231. D. Ohst, D. Pavlik: IEEE Trans. Magn. **25**, 387 (1989)
232. J.H. Price, J.H. Gully, M.D. Driga: IEEE Trans. Magn. **22**, 1690 (1986)
233. R.A. Marshall, W.F. Weldon: Electric Machines and Electromechanics **6** 109 (1981)
234. V. Glukhikh, O. Filatov, V. Belyakov, S. Egorov, V. Kuchinsky, V. Sytnikov, A. Shikov: IEEE Trans. Appl. Supercond. **10**, 771 (2000)
235. S.M. Schoenung, W.R. Meier, W.V. Hassenzahl: IEEE Trans. Magn. **27**, 2324 (1991)
236. D.T. Hackworth, D.W. Deis, P.W. Eckels, D. Marschik: IEEE Trans. Magn. **22**, 1495 (1986)
237. O.K. Mawardi: IEEE Trans. Magn. **18**, 151 (1982)
238. D.E. Johnson, N.D. Clements: IEEE Trans. Magn. **27**, 426 (1991)
239. D.E. Johnson, J.P. Barber, H.L. Laquer: IEEE Trans. Magn. **25**, 266 (1989)
240. A. Pokryvailo, M. Kanter, N. Shaked: IEEE Trans. Magn. **32**, 497 (1996)
241. D.G. Akopyan, Y.P. Batakov, A.M. Dedjurin, A.S. Druzhinin, S.A. Egorov, E.R. Zapretilina, A.I. Kostenko, V.G. Kuchinsky, B.A. Larionov, N.A. Monoszon, G.V. Trokhachev, O.G. Filatov, V.V. Andrianov, N.M. Kolyadin, E.P. Polulyakh, Y.A. Bashkirov, V.E. Ignatov, V.E. Sytnikov: IEEE Trans. Magn. **28**, 398 (1992)

242. J.P. Caron, F.M. Sargos, A. Rezzoug, J. Leveque, D. Netter: IEEE Trans. Magn. **31**, 1480 (1995)
243. P.N. Murgatroyd: IEEE Trans. Magn. **25**, 2670 (1989)
244. S.D. Peck, J.C. Zeigler: IEEE Trans. Magn. **30**, 1639 (1994)
245. C. Polk, R.W. Boom, Y.M. Eyssa: IEEE Trans. Magn. **28**, 478 (1992)
246. G. Schönwetter, C. Magele, K. Preis, C. Paul, W. Renhart, K.R. Richter: IEEE Trans. Magn. **31**, 1940 (1995)
247. N.D. Clements, D.E. Johnson: IEEE Trans. Magn. **27**, 421 (1991)
248. T.L. Francavilla, R.D. Ford, W.H. Lupton, N.McN. Alford, C.S. Saunders: IEEE Trans. Magn. **27**, 1870 (1991)
249. J.P. Barber, D. Newman, R. Ford, R. Klug: IEEE Trans. Magn. **29**, 907 (1993)
250. J.P. Barber: IEEE Trans. Magn. **25**, 68 (1989)
251. J.P. Barber, T.D. Smith, S. Foley: IEEE Trans. Magn. **25**, 75 (1989)
252. E.E. Bowles, W.G. Homeyer, J.M. Rawls, M.W. Heyse, J.B. Cornette, N.E. Taconi Jr.: IEEE Trans. Magn. **29**, 911 (1993)
253. R.D. Ford, N.E. Johnson, F. Cristopher, R. Klug, M. Matyac, N. Lucas: IEEE Trans. Magn. **29**, 943 (1993)
254. E. Aivaliotis, M. Peterhans: IEEE Trans. Magn. **25**, 40 (1989)
255. D.R. Peterson, J.H. Price, J.L. Upshaw, W.F. Weldon, R.C. Zowarka Jr., J.H. Gully, M.L. Spann: IEEE Trans. Magn. **27**, 369 (1991)
256. R.L. Sledge, D.E. Perkins, B.M. Rech: IEEE Trans. Magn. **25**, 519 (1989)
257. E. van Dijk, P. van Gelder: IEEE Trans. Magn. **31**, 84 (1995)
258. J.C. Verite, T. Boucher, A. Comte, C. Delalondre, O. Simonin: IEEE Trans. Magn. **31**, 1843 (1995)
259. A. Pokryvailo, M. Kauter, N. Shaked: IEEE Trans. Magn. **34**, 655 (1998)
260. D.F. Alferov, V.P. Ivanov, V.A. Sidorov: IEEE Trans. Magn. **35**, 323 (1999)
261. J.A. Pappas, S.P. Pish, M.J. Salinas: IEEE Trans. Magn. **35**, 367 (1999)
262. E. Spahn, G. Buderer, W. Wenning: IEEE Trans. Magn. **35**, 378 (1999)
263. N.P. Bublik, Yu.I. Isaenkov, G.D. Kuleshov, P.G. Makeev: IEEE Trans. Magn. **37**, 290 (2001)
264. H.-S. Lee, Y.-S. Jin, J.-S. Kim, C.-H. Cho, G.-H. Rim, J.-S. Kim, J.-H. Chu, J.-W. Jung, V.a. Sidorov, D.F. Alferov: IEEE Trans. Magn. **37**, 371 (2001)
265. T.F. Podlesak, F.M. Simon, S. Schneider: IEEE Trans. Magn. **37**, 385 (2001)
266. H. Singh, C.R. Hummer: IEEE Trans. Magn. **37**, 394 (2001)
267. E. Spahn, G. Buderer, C. Gauthier-Blum: IEEE Trans. Magn. **37**, 398 (2001)
268. H.G. Wisken, F. Podeyn, T.H.G.G. Weise, J. Dorn, D. Westerholt: IEEE Trans. Magn. **37**, 403 (2001)
269. I. Vitkovitsky: *High Power Switching* (van Nostrand Reinhold, New York 1987)
270. D.W. Larson, F.W. MacDougall, X.H. Yang, P.E. Hardy: 'The Impact of High Energy Density Capacitors with Metallized Electrode in Large Capacitor Banks for Nuclear Fusion Applications'. In: *9th IEEE Pulsed Power Conf., Albuquerque, New Mexico, 1993*, ed. by K. Prestwich, W. Baker (IEEE, New York 1993) pp. 461-464
271. T.H.G.G. Weise, S. Jungblut, F.W. MacDougall, X.H. Yang: 'Feuerstellung 2000: Characteristics and Performance of High Eenergy Density Capacitors Applied in the 30 MJ Pulsed Power Supply System'. In: 5th *European Symp. on Electromagnetic Launch Technology*, Toulouse, France, 1995
272. F.W. MacDougall, P.E. Hardy, P. Winsor IV, X.H. Yang: 'The Impact of High Energy Density Capacitors with Metallized Electrodes in Electric Launch Applications'. In: 4th *European Symp. on Electromagnetic Launch Technology*, Celle, Germany, 1993

273. W.J. Sarjeant, D.W. Larson, F.W. MacDougall, I. Kohlberg: IEEE Trans. Magn. **33**, 501 (1997)
274. B. Augsburger, B. Smith, I.R. McNab, Y.G. Chen, D. Edwards, S. Gilbert, G. Savell, M. Robinson, P. Chapman: IEEE Trans. Magn. **31**, 10 (1995)
275. G.L. Bullard, H.B. Sierra-Alcazar, H.L. Lee, J.L. Morris: IEEE Trans. Magn. **25**, 102 (1989)
276. N.H. Fletcher, A.D. Hilton, B.W. Ricketts: J. Phys. D, Appl. Phys. **29**, 253 (1996)
277. L.L. Alston: *HV Technology* (Oxford Press, London 1968)
278. J. Kammermeier: 'Physical and Chemical Conditions for Self-Healing in Metallized Capacitors'. In: *Proc. Symp. on High-Energy Density Capacitors and Dielectric Materials* (NAS/NAE/IOM National Academy Press, Washington D.C. 1981)
279. H.G. Wisken, F. Podeyn, T.H.G.G. Weise: IEEE Trans. Magn. **35**, 394 (1999)
280. J. Yun-Sik, II.-S. Lee, G.-H. Rim, J.-S. Kim, J.-S. Kim, J.-H. Chu, J.-W. Jung, D.-W. Hwang: IEEE Trans. Magn. **37**, 165 (2001)
281. K.M. Slenes, P. Winsor, T. Scholz, M. Hudis: IEEE Trans. Magn. **37**, 324 (2001)
282. H.G. Wisken, F. Podeyn, H.G.G. Weise: IEEE Trans. Magn. **37**, 332 (2001)
283. G.-H. Rim, C.-H. Chu, Y.-W. Choi, Y.-S. Jin, E.P. Pavlov: IEEE Trans. Magn. **37**, 389 (2001)
284. D. Schneider, J. Salge: Z. Angew. Phy. **31**, 346 (1971)
285. General Atomics Energy Products, 4949 Greencraig Lane, San Diego, CA 92123, U.S.A.; Web: www.hvpower.com
286. R. Carruthers: 'The Storage and Transfer of Energy in High Magnetic Fields'. In: *Proc. Int. Conf. on High Magnetic Fields, MIT, Cambridge Massachusets, 1-4 Nov. 1961*, ed. by H. Kolm, B. Lax, F. Bitter, R. Mills (MIT, Cambridge 1961 and John Wiley, New York 1961) pp. 307-318

Index

adiabatic heating, 1
Ampere's law, 3
attenuation function, 133
axial displacement, 125
axial field, 55
axial stress, 89

background field, 77
battery, 131
Biot-Savart, 6, 7, 16, 38
Bitter coil, 9, 35, 43, 51, 60, 81
Boltzmann constant, 29
boundary condition
– cooling, 57
– finite heat capacity, 57
– finite mechanical strength, 57

calculation of magnetic fields, 2
capacitive energy storage, 132
capacitor, 131
coil
– Bitter, 9, 35, 43, 51, 60, 81
– constant current density, 35, 42, 50, 59, 73
– Gaume, 35, 44, 52
– in an external field, 77
– infinitely long, 54
– internal reinforcement, 123
– Kelvin, 35, 45, 52
– nested, 79
– optimal shape, 47
– polyhelix, 36, 48, 89
– resistive, 54
– stress-optimized, 61, 63
– wire-wound, 9, 35, 74, 106, 123
commutating resistor, 154
conductivity, 4
constant current density coil, 35, 59, 73
cooling, 30
copper, 72
– -beryllium, 72, 75
correction factor

– pulse shape, 28
Coulomb gauge, 6
coupling constant, 151
crowbar diode, 145
current density, 28, 37, 89
– characteristic, 57, 58
– distribution, 37, 54
– thermally defined, 58
current density distribution, 89
current pulse, 28
current pulse shape, 58

Debye, 29
– specific heat, 29
– temperature, 29
deformation
– plastic, 1
diffusion
– length, 140
– time, 140
displacement, 3
– axial, 125
– radial, 125
displacement current, 3
distribution function, 55
Dulong and Petit, 29

eddy currents, 1, 106, 131, 140, 143
– open cylinder, 142
elastic deformation, 21
elastic limit, 26
elastic regime, 105, 123
elasticity, theory of, 1, 21
electric current, 3
electric field strength, 3
elliptic integrals, 14
energy storage, 131
equilibrium condition, 1, 56, 74, 91, 104, 106
– solenoid, 25
equilibrium of forces, 24
external field, 61, 73, 77, 140

Index

Fabry factor, 36, 40
– calculation, 40
– second, 41
Fabry formula, 35, 36, 41
– classical, 36
– second, 36, 41
factor
– of pulse shape, 34
– of self inductance, 41
Faraday's law, 3
FEM, 107
field calculation, 89
field strength
– electric, 3
– magnetic, 3
filling factor, 37, 40, 55
final temperature, 28
finite
– cooling capability, 58
– heat capacity, 58
– mechanical strength, 57
finite element method, 107
flux
– magnetic, 3
flux compression, 155
flywheel generators, 131
forces
– radial transmission, 73

gauge transformation, 6
Gaume coil, 35, 44, 52
Gauss's law, 3
Gauss's theorem, 17
Grüneisen, 29

H-field, 3
harmonic analysis, 10
heat
– ohmic, 28
– specific, 28
heating
– adiabatic, 1
– of pulsed coils, 27
helium, 30
homogeneity, 9
homogeneous coil, 50
hoop stress, 89

ideal polyhelix, 104
inductance, 7, 8, 19
– matrix, 89
– mutual, 150
inductive energy transfer, 131, 147
inductor, 131

infinitly high fields, 93
initial temperature, 28
intersection radius, 80

Kelvin coil, 35, 45, 52
– optimal, 46
Kohler plot, 33

Lamé's coefficients, 25
Laplace's equation, 5
Legendre polynomials, 9–11
Lenz's law, 3
Lenz's rule, 140
linearization, 123
liquid
– helium, 30
– nitrogen, 30
Lorentz force, 1, 2, 4, 21, 90, 123

magnetic energy, 7, 56, 68, 89, 96
– density, 7
magnetic field
– along axis of solenoid, 40
– Bitter coil, 19
– calculation, 2
– characteristic, 57, 58
– coil with constant current density, 18
– source-free, 3, 6
– thermally defined, 58
magnetic field strength, 3
magnetic flux, 3, 140
magnetic vector potential, 6, 11
– ring-shaped element, 12
– thin cylinder, 13
– wedge element, 15
magnetoresistance, 1, 33, 106, 131, 136
magnetostatics, 4
maragin steel, 72
material integral, 28, 34, 58
Maxwell's equations, 1, 3
– differential form, 3
– integral form, 3
Mises stress, *see* von Mises stress
mutual inductance, 2, 8, 150, 154

nested coils, 79
nitrogen, 30
NLP problem, 93
normal stress, 24
normalized
– pressure, 76
– radius, 76
numerical optimization, 89

ohmic heat, 28, 56
ohmic power, 56, 70
opening switch, 156
optimization with constraints, 92
outer reinforcement, 73, 81

parallel switching scheme, 156
permeability, 4, 141
permittivity, 4
plastic deformation, 1, 21, 123
plastic regime, 123
Poisson ratio, 23, 24, 73, 75, 107
polyhelix, 104
polyhelix coil, 36, 48, 89
pulse length, 28, 34, 58, 69
pulse shape factor, 34
pulsed coil, 5
– cooling, 30
– heating, 27

quasi-stationary, 1
quasi-stationary fields, 4

radial displacement, 125
radial transmission of forces, 73
residual resistivity ratio, 30
resistivity, 28, 29, 141
RLC circuit, 132
RRR value, 30
Runge-Kutta method, 134, 135, 137

S2-glass, 123, 127
scalar magnetic potential, 5, 9
self-inductance, 8, 151
– factor, 20–22, 41
– solenoid, 20
separation of layers, 124
serial switching scheme, 158
skin depth, 141
solenoid, 9
SolvOpt, 93, 96
source-free magnetic field, 3, 6
spatial homogeneity, 9
specific heat, 28, 29
strain
– tensor, 1
– von Mises, 123

strain tensor, 23
strength theory, 26
stress
– axial, 89
– hoop, 89
– normal component, 23
– tangential component, 23
– tensor, 1, 24, 91, 106
– thermal, 1
– von Mises, 1, 27, 75, 91, 123
stress tensor, 1, 23, 24, 56, 74, 91, 106
stress-optimized coil, 61, 63
stress-strain
– curve, 26
– relation, 25, 56, 123
superposition principle, 4
surface integral, 16

tangential stress, 24
Teflon, 124
temperature
– final, 28
– initial, 28
tensor
– strain, 1
– stress, 1, 24, 91, 106
theory of elasticity, 1, 21
thermal stress, 1
time constant, 140

vector potential, 89
volume integral, 16
von Mises
– strain, 123
– stress, 1, 27, 75, 89, 91, 123
– tensor, 1
– yield criterion, 27

wire-wound coil, 9, 35, 42, 74, 106, 123

yield criterion
– von Mises, 27
Young's modulus, 22, 24, 73, 107, 123

zonal harmonic analysis, 10, 11
Zylon, 123, 125

Druck: Strauss Offsetdruck, Mörlenbach
Verarbeitung: Schäffer, Grünstadt